The Odyssey

of KP2

The Odyssey

of KP2

AN ORPHAN SEAL, A MARINE BIOLOGIST,

AND THE FIGHT TO SAVE A SPECIES

Terrie M. Williams

THE PENGUIN PRESS

New York

2012

THE PENGUIN PRESS
Published by the Penguin Group
Penguin Group (USA) Inc., 375 Hudson Street, New York, New York 10014, U.S.A. • Penguin Group
(Canada), 90 Eglinton Avenue East, Suite 700, Toronto, Ontario, Canada M4P 2Y3 (a division of Pearson
Penguin Canada Inc.) • Penguin Books Ltd, 80 Strand, London WC2R 0RL, England • Penguin Ireland,
25 St. Stephen's Green, Dublin 2, Ireland (a division of Penguin Books Ltd) • Penguin Books Australia Ltd,
250 Camberwell Road, Camberwell, Victoria 3124, Australia (a division of Pearson Australia Group Pty Ltd) •
Penguin Books India Pvt Ltd, 11 Community Centre, Panchsheel Park, New Delhi – 110 017, India •
Penguin Group (NZ), 67 Apollo Drive, Rosedale, Auckland 0632, New Zealand (a division
of Pearson New Zealand Ltd) • Penguin Books (South Africa) (Pty) Ltd,
24 Sturdee Avenue, Rosebank, Johannesburg 2196, South Africa

Penguin Books Ltd, Registered Offices:
80 Strand, London WC2R 0RL, England

First published in 2012 by The Penguin Press,
a member of Penguin Group (USA) Inc.

Image credits
Pages ii, 65: David Williams (NMFS permit no. 13602-01, Terrie M. Williams)
1: Teri Rowles (NMFS permit no. 932-1905, Teri Rowles)
127: National Marine Fisheries Service
133: Terrie M. Williams
266: Patricia Sullivan

Map illustration by Meighan Cavanaugh

LIBRARY OF CONGRESS CATALOGING IN PUBLICATION DATA
Williams, Terrie M.
The odyssey of KP2 : an orphan seal, a marine biologist, and the fight to save a species / Terrie M. Williams.
p.   cm.
Includes index.
ISBN 978-1-59420-339-8
1. Hawaiian monk seal—Conservation.   2. Wildlife conservation—Hawaii.   3. Wildlife rehabilitation—
Hawaii.   4. Endangered species—Hawaii.   5. Williams, Terrie M.   I. Title.
QL737.P64W55   2012
599.79′23—dc23   2011050415

Printed in the United States of America
1   3   5   7   9   10   8   6   4   2

Designed by Meighan Cavanaugh

To the Volunteers:

the heart of the ocean

One touch of nature makes the whole world kin.

—*William Shakespeare*

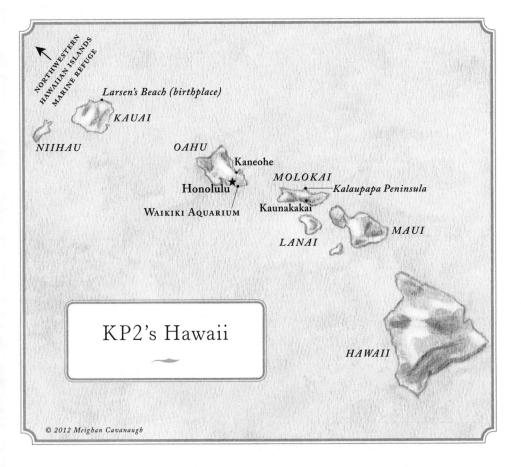

NORTHWESTERN HAWAIIAN ISLANDS MARINE REFUGE

Larsen's Beach *(birthplace)*

KAUAI

NIIHAU

OAHU

Kaneohe

MOLOKAI

Honolulu

Kalaupapa Peninsula

Waikiki Aquarium

Kaunakakai

LANAI

MAUI

KP2's Hawaii

HAWAII

© 2012 Meighan Cavanaugh

Contents

PART III

Survival

Preface

In a violent spring downpour off the southern coast of Molokai, an outrigger canoe threatened to capsize in rough waters. Six paddlers were buffeted in the seas as they strained to reach the protective harbor in Kaunakakai. Seawater and rain poured into the canoe, mixing with the sweat of the struggling team. As dark clouds descended and with their heads bent in exertion, the paddlers soon lost their bearings. Small and exposed, only the thin shell of the craft separated them from tiger sharks patrolling the waters below.

Impending darkness and the chill of hypothermia intensified the danger, when suddenly a small bald head popped out of the water. Spiky eyebrow hairs gave the smooth gray head a comical, monkish appearance. The head floated in the distance and then sank among the waves. Before any of the paddlers could react, the young Hawaiian monk seal reappeared alongside their canoe.

"KP2!" one of the paddlers shouted. The ten-month-old monk seal had arrived at Kaunakakai Wharf several weeks before and had taken to swimming with the local children.

He was unlike any monk seal or ocean creature the residents of Molokai had ever seen. Instead of lolling on beaches like most seals, KP2 shivered with energy as he bounced around the harbor greeting people as if he were one of the family. He called to children with an enthusiastic flipper slap to the water and swam circles around their treading legs, eliciting squeals of delight. When the mother of one of the children waded into the water with a pink boogie board, the locals soon discovered that this young seal also had a unique talent. Somehow, somewhere, KP2 had learned to surf. Upon seeing the boogie board, he immediately flopped his body on top and propelled himself across the water with his hind flippers churning up a white frothy wake.

The island community was left stunned by this unusual, friendly seal. Considering him a gift from the ocean, islanders extended their highest honor to KP2, due to his obvious love of people and his surfing skill: they now called him "local."

On this particular stormy evening the curious seal had followed the paddlers out beyond the wharf. Oblivious to the downpour, the seal attempted to board the canoe. He propped his white chin on the gunnel seeking a playmate, further destabilizing the boat.

"KP2, go home!" the paddlers yelled.

The young seal slipped dutifully back into the water, waiting for someone to follow him as they usually did. When no one jumped in, he sensed that something was different. KP2 circled the outrigger canoe once more and began to swim along the tops of the waves. The canoe team paddled in earnest to keep up with the seal, relying on his innate navigational skill. Rather than diving, which would have made his travel much easier, the seal surfed the waves and periodically turned back to watch the progress of the paddlers behind him.

Soon the jutting outline of the wharf and the lights of town were visible. Within twenty minutes the grateful team and their monk seal escort were slipping into the protected waters of Kaunakakai Harbor. With the canoe tied up and the paddlers heading home to recover from the chill, KP2 swam off. Instead of heading out to sea like a normal wild seal, he climbed onto the back transom of a docked sport-fishing boat to await the return of children in the morning and the resumption of their water play.

Unbeknownst to the snoozing seal, a series of events would soon drive a distance between him and the island people he'd befriended. He was destined to undertake a journey unlike anything experienced by his species. Years would pass, yet KP2 would never forget Hawaii or the children of the islands. They in turn would not forget him.

Most wild seals live their entire lives without a name. Mothers easily recognize their pups just from the pitch of their cries. They communicate with each other and the rest of the ocean's inhabitants through sounds and smells and actions. Hungry seals eagerly consume unidentified fish without formal introductions.

The anonymity of nature would suddenly change with the arrival of KP2. This remarkable seal, also lovingly known as Ho'ailona, Smoodgey, and Mr. Hoa, inspired nicknames that were as colorful as his personality. Over the years, he would be called Butthead, Honey Boy, Fish Stealer, Little Angel, Bugger, and Elvis of the Seals, depending on his mood and his audience.

His celebrity began with his fascination with people, a behavioral anomaly among seals that would, aside from creating a remarkable extended family, repeatedly land him in trouble. Wild animals are not supposed to act this way—every fiber of undomesticated creatures urges them to escape human presence. Yet on very rare occasions, for reasons scientists cannot fully explain, an animal breaks from the pack to join us. Such was the case with KP2.

He was a wild seal, a member of an endangered species who left his own world to play in ours. Ancient Hawaiians called his kind *ʻilio holoikauaua*—"the dog that runs in rough water." I simply called him *hoa*—"friend."

From the moment of his birth, he changed how many of us view our lives on this fragile planet.

The Odyssey
of KP2

PART I

Destiny

1.

Birth

My earliest aspiration, at five years old, was to grow up to be a dog. This seemed the noblest of professions, and I determined that it was only a matter of time and desire before I grew the requisite four legs and tail. My religious parents, however, had equally unlikely expectations and prayed that I'd become a Roman Catholic nun.

Sister Everista and Sister Agnes never knew their true influence on the girl known best for scraped-up knees and a love of the outdoors. Instead of civilizing the animal out of me, the stern-faced, black-habited sisters inadvertently taught me how to communicate with the "lowly creatures" of the Bible.

I found that I could perceive the nearly invisible body and eye movements comprising animal language, and predict an animal's next move as if I were inside its mind. It wasn't communication in a Dr. Dolittle sense; rather, I was able to "read" the local dogs, cats, foxes, squirrels, and rabbits as others might read a newspaper.

At every opportunity I'd escape the disinfected halls of the parochial schools and plunge into the wild chaos of the surrounding oak forests of the East Coast. The freedom to poke around creeks like an otter in search of frogs or to slip through thorny blackberry bushes with the liquid movement of a fox was exhilarating. My animal senses grew with time, much to the consternation of the nuns and the rest of the girls in my class. I was known somewhat disparagingly as "that girl who likes animals." Trips to the confessional enforced by Sister Agnes with a twisting two-fingered clamp on my ear were, more often than not, to confess to the sin of having released some rescued frog, baby bird, or field mouse that had wriggled free of my pocket and crawled between the church pews. I considered the litany of Hail Marys recited on scabbed knees due penance for the creature's salvation.

Over the years this fascination with the furred and finned evolved into a lifetime of globe-trotting in the man's world of wildlife research. I knew that my success had less to do with raw intelligence and more with an innate ability to relate to animals. If I couldn't *be* an animal, then at least I could learn to appreciate the intimate details of their daily lives by studying them. The wilderness became my cathedral. Skittish cheetahs, playful dolphins, mitten-pawed sea otters, and stoic Antarctic seals were my congregation. Dominican discipline taught me focus; Mother Superior's demands for self-sacrifice honed an inborn skill for animal empathy.

Yet in all my wildlife encounters encompassing a lifetime of adventures, there was one major disappointment. No wild animal had ever read me in return. A Pembroke Welsh corgi named Austin, the canine member of my tiny Hawaiian 'ohana (family), had come the closest. But when it came to the inhabitants of the woods and the oceans, animal communication had been a lonely, one-way affair. Nuns and scientific textbooks espoused that such was the nature of animals, since nonhuman creatures possessed neither souls nor intellect. Here I must say that both were mistaken. For, unexpectedly, after half a century of being the mind reader, one of them suddenly read me.

He was not the fastest, biggest, or purportedly smartest of animals; rather he was an immature, nearly blind sea mammal that had been cast out by his own species. By all rights he should have died on an isolated beach on Kauai. Like me, he began life attempting to cross physical and societal boundaries that separated humans from animals, oblivious to the impossibility. He was a boisterous surprise in a scientific career that was in danger of maturing into comfortable cynicism.

KP2 CAME INTO THE WORLD on May 1, 2008, in the usual way of seals—slippery, wet, and sliding unceremoniously from between his mother's back flippers onto a nursing beach. With a shake of a head covered in thick black lanugo, the dark fetal fur jammies of his species, he opened his eyes to tropical tranquillity while resting a chin on scattered fragments of bleached coral. In stark contrast to the harsh, icy landings endured by his cousin Antarctic seal species, KP2's birthplace on North Larsen's Beach, Kauai, was one of turquoise blue water warmed by an equatorial sun. In the distance, wisps of steam rose from lush vegetation blanketing the towering *pali*, the mountainous peaks protecting the beach from the afternoon trade winds.

The newborn seal stretched, uncoiling flippers that had been wrapped snugly around his body in utero for the past ten months. Carried within the belly of his mother, RK22 (her National Marine Fisheries Service ID), the developing pup had submerged hundreds of feet in depth in pursuit of fish that would sustain his fetal growth. For nearly a year he had been rocked to sleep on the tides of the Hawaiian current, cradled in his mother's womb as she swam back and forth between the islands, coral atolls, and nursing beaches.

THE QUIET OF KP2'S ENTRANCE into the world was broken within hours of his first breaths by an explosion of violent splashing immediately offshore. Raucous calls of brawling seals had awoken him and made him cry out. A throaty growl soon overwhelmed his high-pitched calls for his mother as the seal pup was attacked. In a single blow, an adult male Hawaiian monk seal nipped and rolled the newborn seal into the sand. The aggressive male left deep gouges along the tide line as he maneuvered for a second, more lethal hold on KP2's neck—a hold that could easily have broken the pup in two.

KP2's attacker was just as likely as not his father. Lumbering heavily on powerful front flippers and a massive chest, the four-hundred-pound male dwarfed KP2. He would have crushed the newborn had KP2 not scrambled out of the way. The male had only one goal on his testosterone-driven mind: to mate with RK22. Her pup was merely a rock-sized obstacle in his path.

Inexplicably, KP2's mother did nothing. As her pup tried to escape among the jagged lava rocks, she watched passively. It was uncharacteristic behavior for a monk seal mother, or any seal mother, for that matter. Among the many species of pinnipeds, the collective name for the mammalian group that includes sluglike phocid seals and clownish-eared sea lions, male aggression toward newborns is not unusual. But seal mothers, even in cases in which they are outweighed three to one by full-grown males, will—even within minutes of giving birth—reproach such aggressors with teeth bared, ready to sink them into and draw blood from old, scarred chests. On nursing beaches from the tropics to the polar sea ice, pinniped mothers risk death to defend their helpless pups.

RK22 showed no such courage against her pup's aggressor. Rather, she ignored the ruckus as KP2 was mauled, seemingly irritated by the disturbance to her afternoon nap. Confirming her indifference, RK22 followed the brutish male and another male companion into the water, leaving her ruffled offspring abandoned and suckling on beach rocks for comfort.

"*RRRAUGHH*," KP2 CALLED desperately for his mother. His second day of life was not going much better than the first, and would end just as tumultuously. With hunger overtaking KP2's fear of further attack, he began to call loudly for his mother and her milk. Unlike the charming chirp of sea otter pups or the high-pitched squealing whistles of dolphin calves, KP2's cry was a distinctive rumbling "*rrrraaughhh*."

Each rough-edged bawl of the pup contained a vocal signature that could be distinguished by his mother from any other seal's call, had there been others around. An unbreakable mother-pup bond existed between KP2 and RK22 that transcended her lack of interest. A permanent connection, stronger than the pull of the moon on the tides, had formed deep within the instinctual parts of their brains the instant he had wriggled free from her placenta. Nose to nose and whisker to whisker, they had sniffed and vocalized in pinniped recognition of each other's earthly existence in those first moments. Their mutual greeting in the shared instant of birth had sealed their genetic knot. She could not deny that he was biologically tied to her or that his calls were beckoning her.

When KP2's mother finally responded to his cries, she was accompanied by the large, pugnacious male who trailed persistently behind her. Again KP2 was attacked. But this time, it was his own mother who turned on him. With an aggression typically reserved for territorial fights, she took his head into her mouth and bit down. If she had wanted, she could have killed her offspring instantly by puncturing his skull with her sharp canine teeth. Instead, she grabbed the struggling pup in her mouth, shook him from side to side like a dog with a wet dish towel, and spat him out on the ground. KP2 rolled in a limp bundle of matted fur and sand, his uncoordinated flippers flailing in the ocean debris of the high-tide line.

More flustered than hurt, the pup struggled to regain his footing. Disoriented and battered, KP2 was not sure which way to crawl. So he hunkered down in the sand where he had landed.

KP2's STRUGGLES had been witnessed by a group of islanders who were well aware of RK22's maternal incompetence. Shadowed beneath hats and hidden behind the tropical vegetation, they tried to blend into the background as they watched with binoculars so as not to disturb the mother and her new pup. The previous June, less than a hundred feet from the bloodstained sand that marked KP2's birthplace, this same group had witnessed the entry of his sister into the world.

To these members of the National Oceanic and Atmospheric Administration (NOAA) Kauai Monk Seal Team, this adult female seal was notorious for bad behavior and bad mothering. She had been just as unresponsive to the cries of KP2's sister. The year before, the team had waited three days before they could no longer stand to watch idly as the tiny female pup slowly, helplessly starved to death on the beach. Taking pity on the abandoned pup, they finally picked her up.

For days the dedicated team tried valiantly to reunite KP2's sister with her mother by moving the crying pup directly in front of RK22's sunbathing spot. But RK22 would have none of it. Instead she snoozed in the sun, ignoring the persistent calls of her pup. When the newborn seal became too loud, her mother entered the water for a swim and in a single dive drowned out the sound of her starving offspring.

Finally admitting defeat in the face of RK22's indifference, one of the team members called in a veterinarian, who humanely euthanized the weak and emaciated pup. KP2's sister lived for only five days.

Determined not to let KP2 suffer the same fate, the Kauai team called the marine mammal stranding headquarters on the neighboring island of Oahu. David Schofield of the National Marine Fisheries Service Pacific Islands Regional Office (NMFS-PIRO) answered. With an office in downtown Honolulu, David maintained an uneasy relationship with the local Hawaiians and at times his upper management. He was the federal official in charge of marine mammal strandings for the islands, and unlike his bosses 4,519 miles away in Washington, D.C., David's everyday decisions were entangled with cultural sensitivities as well as ocean politics. While Washington bureaucrats could ponder their next moves, David's job demanded instantaneous life-or-death evaluations for the dolphins, whales, and seals that found their way onto Hawaii's shores. It was simply impossible to please everyone.

Having grown up in the shadow of the Trump casinos in Atlantic City, a brusque decision-making attitude came naturally to David. However, his East Coast edge for dealing quickly and independently with stranded animals in Hawaii sometimes chafed island sensitivities. Aloha shirts, a bicep tattoo, and a shared passion for surfing and canoeing the waters of Waikiki notwithstanding, David remained a *haole* (outsider).

Issuing a series of rapid-fire phone calls, David dealt with KP2, aware that some of his decisions would not be popular; in these situations it was easier to ask for forgiveness than for permission. As a result, less than forty-eight hours after his birth, KP2 was whisked away from his abusive mother, scooped up fireman style by David and another government official, Shawn Farry, from the Pacific Islands Fisheries Science Center, or PIFSC, a unit of the National Marine Fisheries Service. David grasped the little seal's front flippers and Shawn had hold of his back flippers while his thick pink umbilical cord dangled between them. Either man could have easily carried the seal pup, at only thirty-four pounds, under one arm. Instead they gently cradled the stunned seal between them and placed him in a dog kennel.

A quick inspection of KP2 showed that he was no worse for wear despite his postpartum ordeal on the beach. He had been roughed up but there was no evidence of open bite wounds. From what David and Shawn could see, KP2 was a robust, healthy monk seal pup. "So why," they asked each other, "did RK22 abandon him?"

Nature's law dictating survival of the fittest initially seemed the most likely explanation. Because the metabolic cost of pregnancy is low compared to lactation, many mammals (including those that live in the sea) are known for carrying fetuses to term only to abandon them at the moment of birth. Reasons for this seemingly wasteful behavior vary. Sometimes mothers recognize that their newborns are ill or otherwise deformed. Other times mothers know that local food resources are insufficient for sustaining both their own caloric needs and those of their young. Nature dictates that, above all else, you save yourself. This is the harsh biological reality for wild mothers caught in rapidly degrading habitats: cut your losses and focus efforts on the hope of guaranteeing the survival of next year's healthy offspring.

But the pup in question was not ill, nor had RK22 disappeared to forage elsewhere. KP2's mother simply appeared to prefer male companionship to motherhood. Consequently, her offspring had been left to find their own path to survival.

David and Shawn now faced the most difficult decision encountered by a biologist. They could let nature run its course, or they could intervene. With so few Hawaiian monk seals left in the world, the decision was obvious. They chose to intervene, and in a remarkable turn of events KP2 was given a second chance at life.

AT THE NEARBY LIHUE AIRPORT, a U.S. Coast Guard C-130 transporter waited next to a line of palm trees. David Schofield had arranged a deal for the pup to hitch a free ride on the aircraft as it returned to Oahu. The C-130's engines roared in anticipation of takeoff as the vehicle containing KP2 and his rescuers sped across the tarmac. The seal stuck his nose through the grating and peered through the small holes of the kennel. He had no perception of what the normal sights, sounds, and smells of a monk seal's life should be. However, despite the noise and urgency surrounding him, there was some comfort in not being mauled or abused by one of his own.

With a vibrating roar of propellers, the seal pup left behind his mother to her brutish male companion in the turquoise waters of Kauai. It was an unprecedented act instigated by a bad mother, organized by a brash ex–New Jerseyan, and executed by the generosity of the U.S. Coast Guard.

EVENTUALLY, THE LITTLE BLACK PUP who'd been abandoned on a lonely Kauai beach would become an icon, a rogue, and a threat. Unbeknownst to RK22, the survival of her entire species would one day come to rest on her offspring's small and once discarded shoulders.

2.

Explorations

Eight of us were on the verge of wholesale cabin fever holed up in McMurdo Station, Antarctica, eight hundred miles from the South Pole, as the coldest September on record chilled our Weddell seal expedition to an icy halt. Although I had weathered seven expeditions to the frozen continent, September 2009 was by far the most brutal. At −70°F, the conditions outdoors were life-threatening. Winds forced the thermometer to −100°F, and nothing moved. The oil in our snow machines had long frozen to the viscosity of mud, while the moisture from our breath stuck to our eyelashes and froze our eyes shut.

My team was slowly recovering from the fallout of a close call on the sea ice. Four of us had been caught in a whiteout while trying to locate Shaquille O'Seal. The Weddell seal had been missing for weeks, hiding beneath snow-bridged cracks and leading us on a wild chase across McMurdo Sound. All of the seals we were studying had taken shelter in the comparatively warm waters below the sea ice. Unable to follow, we were left shivering in the wind and blowing snow, marveling at how the slushy 27°F ice water steamed like a thermal spa in the seals' disappearing wake.

Shaquille was one of several seals each carrying over $50,000 of scientific instrumentation from the National Science Foundation in a backpack. A miniature video camera and computerized sensors in his pack were designed to record the intimate details of his underwater exploits. Through the instrumentation we were able to dive virtually into the Weddell seal's icy world. We monitored every swimming stroke and beat of their hearts as the seals descended to extraordinary depths. Sometimes this meant distances of more than a third of a mile from a breathing hole on the icy surface to the benthic zone, the darkest bottom of the ocean. In the cold and dark and under intense hydrostatic pressures, these master divers stalked giant four-foot Antarctic cod. Through our miniature instruments, my expedition members and I were privileged to have a ringside seat as predator and prey fought for survival in the harshest environment on earth.

Our diving seals had revealed incredible polar sights never before witnessed by human divers, from secret ice passages beneath coastal glaciers to gardens composed of alien-looking plants growing five hundred meters below the sea ice. Many expeditions before ours had explored the surface of Antarctica; we thrilled to the discovery of what happened below the ice.

The instrumented seals in our study were free to swim and dive throughout McMurdo Sound. Usually, after a week of hunting, the seals hauled out on top of the ice to rest, giving my team the opportunity to retrieve the backpack and all of our data. But the onset of foul weather had altered Shaquille O'Seal's pattern, causing the seal to swim off with our instruments and our science. Weeks passed without a sighting, leaving my expedition members anxious and on edge.

Bone-chilling surveys on snowmobiles crisscrossing the sea ice in the hope of spotting a sign from the seal left all of us nearly hypothermic. We wrenched spines and snow machines on the rough sastrugi covering the frozen landscape. When we turned to high tech by listening for radio and satellite signals from the tags on Shaquille's backpack, we were continually disappointed. The radio receivers remained ominously silent. Our computers never woke to a satellite hit.

After nearly a month of waiting, I feared the worst for the missing seal and our instruments. We needed a new plan. I sought out Traci Kendall and Beau Richter, two marine mammal trainers at the heart of my lab at the University of California, Santa Cruz. Following in the footsteps of polar explorer Ernest Shackleton, who employed animal handlers on his 1914 trans-Antarctic *Endurance* expedition, I found the addition of people experienced in animal behavior invaluable on my research team. Importantly, Beau, Traci, and I shared the gift of being able to read animals. Each of us, in different ways, was driven to use that ability to help preserve the wild animals of the land and seas.

The previous year Traci had been my expedition animal handler and demonstrated a remarkable calming influence on wild Weddell seals. Despite standing at only five foot three in insulated Bunny boots, she had no fear walking up to enormous male seals that were seven times her weight. The giant seals would rear up, snap their jaws threateningly, and try to roll away. The trainer always stood her ground and never flinched. Instead, by moving slowly and always letting the seal know where she was—much like she would act around a skittish horse—Traci calmed the seals to the point that the rest of us could deploy and remove our instrumentation with ease. This year it was Beau's turn as the animal handler on ice, while Traci stayed at home to take care of our dogs and the animals in my lab.

Traci agreed with Beau and me—over a very long-distance phone call—about how we should handle the search for Shaquille. We needed to understand the internal pressures and the instinctual biological drivers that defined his species. Rather than just the challenges we faced as humans in Antarctica, we also had to consider the external pressures—the combination of weather, habitat, and seal society—that comprised the environment in which the seal competed in order to survive. By seeing the world through Shaquille's eyes, we could understand the seal's decisions and anticipate his movements. Only then would we find him.

After consulting with Beau and Traci, I came to the conclusion that from Shaquille's perspective the world was all about size and breathing space. He was exceptionally broad in the shoulders, and had used his brawn to his advantage in finding mates. As one of the largest seals, he could afford to be stubborn, curious, and lazy, as few dared to challenge him due to the confidence of mass. He bossed around the local females and younger males without compunction.

But there was a handicap inherent in this top seal position. Shaquille was so bulky that he could not fit his body through the small cracks and holes in the sea ice used by the smaller seals to breathe and escape the water. Like all Weddell seals, he could hold his breath for over an hour; any longer and he had to find a hole in the ice to grab a breath of air before he drowned. Sooner or later—preferably before twenty-three minutes, at which point he would begin to feel the fatiguing burn of lactic acid in his muscles—he had to surface to breathe.

Severe weather had trapped the massive Shaquille below the ice by freezing over the open water leads that the seals used during the summer to surface and breathe. With few options remaining, we knew that Shaq had to be nearby, floating below one of the tidal cracks to catch his breath. Smaller male and female seals could squeeze periodically through these rare breathing holes and tidal cracks. Hauling out on top of the frozen sea, they basked in the warmth of the rising late winter sun, even if its thin rays were available for only a few hours each day. There was no such reprieve for Shaquille. The seal had resigned himself to staying underwater, too lazy to use his sharp canine teeth to widen breathing holes, or perhaps too savvy to waste his time on such a formidable task. Instead, he muscled his way into another seal's breathing spot. By poking his big snout through the ice opening, he inhaled and exhaled below the ice, huffing for hours where neither the weather nor our team could reach him.

His strategy worked, except for one nagging problem. There was an entire colony of displaced seals holding their breath below him. With lungs near bursting, the submerged seals eventually bit at Shaquille's dangling flippers and forced him out of their way so they could snatch a quick breath of much-needed air.

Based on our assessment of Shaquille's personality and predicament, there was a good chance that we would be waiting a long, long time before he was motivated to move. Something had to change.

"We need a sucker hole," I concluded. "What if we created a breathing hole so big and so attractive that even a big, lazy, territorial seal like Shaq would find it too irresistible not to haul out?"

Beau and Traci agreed, and I presented the idea to the rest of the expedition members. Although skeptical at first, the others warmed to the idea more out of boredom than for its brilliance. The next day we used a four-foot-wide augur to drill a hole though nine feet of solid sea ice in one of Shaquille's favorite hangouts. To make it even easier for the half-ton seal, Beau and I made a series of ice stairs by chipping away the edges of the hole with an ice ax. The rotund seal could slide effortlessly out of our custom-designed breathing hole with little more energy than rolling over.

We didn't have to wait long. In less than twenty-four hours, curiosity got the better of Shaquille O'Seal and we were ready for him. Four of us watched from a distance as our missing seal poked his whiskered muzzle through the slushy ice water. Steam rose several feet above his exposed nostrils, and seawater instantly froze on his whiskers and eyebrows. Surveying the horizon with a 360 degree piroutte in the hole, Shaquille seemed perplexed by the luck of finding such an oasis in the sea ice. Using our icy exit steps, the seal flopped his chest and then his belly forward, hauling his blubbery body out into the open.

"Come on," I whispered under my breath as Beau and two others watched intently with binoculars. We were hidden by snowmobiles several hundred feet away. No one stirred.

"Not yet," Beau cautioned. If we moved too quickly, Shaquille would slip backward into the water before we could reach him. We held our steaming breaths, waiting for Shaquille's back flipper tips to clear the water. The seal inched another several feet forward. We had him.

"Now!" Beau instructed. The group jumped into action on the snowmobiles. With practiced skill, we positioned our vehicles between the seal and the sucker hole. Using the same calming head and eye cover technique as Traci's, Beau soon had the seal quietly lying on his stomach. Before Shaquille had time to raise a flipper in protest, we had his instrument backpack detached.

I smiled at our good fortune as we packed the snowmobiles to leave. The large Weddell seal had already turned over to go back to sleep. Clearly, we were a minor disturbance in Shaq's day compared to the flipper-biting seals he had endured for weeks. I noticed that the sea ice was stained pink with Shaquille's blood from the injuries the other seals had inflicted. As a result of his obstinacy, the huge seal's hind flippers had taken on the same ragged appearance as the wind-torn flags that lined our snowmobile trails on the sea ice. We had done him a great favor by creating the giant hole in the ice. I also realized, however, that in picking that moment for his return, my companions and I had made a potentially fatal error.

"The wind's come up!" I cursed as I looked down at my boots. Although our science was saved, my team had momentarily ignored its own survival. With all of us focused on catching Shaquille, we had missed the telltale drift of sugary ground snow swirling around our heavy, insulated boots. We had also ignored the rising of a South Pole wind that now whistled through our parkas. Suddenly four of us were blinded, stopped dead in our tracks as a wall of snow enveloped our field party.

A charging wind rushed unchallenged across the sea ice where we stood, with gusts of seventy-five miles an hour overtaking us. The rush of snow deafened and muted our team. Every breath was choked with snow. Fine ice crystals shot by each gust cut into our clothes and skin with the viciousness of glass. We were whipped mercilessly by stinging snow for our mistake, and I began to wish that I could follow Shaquille as he slipped back down our sucker hole and into the safety of the calm waters below.

Despite the ground snow, if I looked straight up I could make out blue sky above us and the tips of the surrounding mountains. A peculiar feature of Antarctic snow is that it usually arrives on the back of the wind rather than falling down from clouds in the sky. Through the gray I could see the volcanic smoking cap of Mount Erebus, and at times the distinctive brown outline of Castle Rock. We had no true horizon, but we did have a direction and a handheld GPS.

Together my shivering team quickly agreed on a plan. We created a train of four snowmobiles and began to head toward an intersecting flagged road that we knew led to McMurdo Station. We also agreed that if we did not locate one of the tattered red road flags within twenty minutes we would stop and set up survival tents until the snowstorm cleared. We would be cold, but at least not lost or dead.

Inching our way forward, the four of us headed due west, confident that we would eventually cross the flagged route back to the station. The biggest danger was in losing our bearings and encountering one of the larger cracks in the sea ice. Years before a colleague of ours had died in this area of McMurdo Sound when his track vehicle had become trapped in a crack and then fallen through the ice, dragging him to a dark, cold, watery grave nine hundred feet below us.

Fine snow began to creep into my pockets and between my gloves and parka sleeves. I could feel my left cheek stinging in the cold and knew instinctively that frostbite was not far off if I didn't get the area covered quickly. A brief stop to regroup, look skyward, and check our bearings allowed me to clear my goggles and readjust my clothing. Then we all saw it: a red flag whipping on the end of a wind-bowed bamboo pole. When the team finally reached it, we saw another flag drifting in and out of view in the swirling snow. We slowly drove forward, and then saw another. And then another. Miraculously, we had navigated through the blinding snow to Highway 1, the ice trail to McMurdo Station. By hopping from flag to flag we crawled as a team back to the safety and warmth of the field station.

The next day the storm cleared, leaving a blistered triangle of frostbitten skin on my cheek as a lesson. With the clearing came a chilling stillness that forced the entire team to hole up at the station. The silence was eerier than the roaring wind. Soon expedition isolation, loneliness, and friction set in among the silent, inactive expedition members.

No planes with supplies or "freshies" (fresh fruits and vegetables) would arrive for another month. The same faces encountered at meals grew scraggier by the day. The same monotonous institutional brown foods were eaten. There was little to do but wait out the tail end of winter harshness to continue our scientific research.

E-MAIL BLESSEDLY PROVIDED our one escape to the outside world.

"How would you like a Hawaiian monk seal?" I mused one day, trying to raise a smile out of Beau Richter.

Beau barely looked up from the cup of coffee he was hovering over. Having worked as a dolphin trainer on the island of Oahu in Hawaii, and more recently in my marine mammal physiology lab at the university, Beau rarely showed up without a cup of Kona coffee and sporting flowered board shorts and flip-flops. Even on the expedition he wore Hawaiian gear beneath his insulated pants and parka. He is the only person I've ever known to wear aloha gear in the field on Ross Island, Antarctica. He had shaved off most of his beard and sandy brown hair for the expedition, leaving telltale suntan lines that quickly faded in the Antarctic winter darkness. He was the requisite young brawn on the expedition, and he grew testy with the slow pace.

Beau slowly grinned his first real grin in days. "Sure. Go for it," he responded absently.

"Be careful what you ask for," I warned. The Hawaiian monk seal (scientific name *Monachus schauinslandi*) was the most endangered marine mammal in U.S. waters. Whereas other marine mammals migrated across entire ocean basins, the watery territory of this seal consisted of a small, thin scar on maps of the Hawaiian island chain. It was because of this rarity and isolation that the species was so intriguing. I wanted to use my scientific talents to help. However, these characteristics also made the Hawaiian monk seal nearly impossible to study.

The e-mail on my computer screen had come from the National Marine Fisheries Service headquarters in Maryland. It was short and cryptic, a simple inquiry. "Would you be interested in temporarily caring for an orphaned Hawaiian monk seal pup at your lab?" Jennifer Skidmore, a fisheries management specialist with the Office of Protected Resources (OPR) at NMFS was exploring options. Trained as an ecologist who studied the effects of invasive species on marine habitats, Jennifer was now in charge of finding homes for sick, injured, or otherwise abandoned marine mammals. She was an animal's last chance before euthanasia.

I was both surprised and suspicious. To say that marine mammal researchers and government officials in the NMFS Office of Protected Resources were often enemies would not be far from the truth. Over the years I had weathered numerous skirmishes with the OPR. Admittedly, both sides had the best interests of dolphins, whales, seals, and sea lions in mind. However, field biologists like me, risking our lives and working with marine mammals, often felt that Washington bureaucrats were out of touch with the reality of what was happening to the ocean's animals. Conversely, NMFS officials recognized that researchers were mostly ignorant, often purposely so, of the laws governing marine mammal protection.

Both researchers and NMFS personnel believed in the conservation of the ocean's largest animals; we just took very different approaches to accomplishing that mission.

Jennifer's office issued the research permits that allowed scientists like me to touch marine mammals. Only permit holders could legally approach within five hundred feet of a wild cetacean or pinniped. No permit, no animal access, no research. Researchers often navigated through years of paper bureaucracy and public scrutiny in the quest of obtaining the letter granting permission to study and save marine mammals.

For fifteen years my permit applications to work with Hawaiian monk seals had been denied, and I'd grown cynical about the ability of humans to save any species, much less this critically endangered seal. I recognized that my permit denials had little to do with science and everything to do with politics. From the government side, one did not take chances with an endangered species, regardless of a scientist's good intentions. There were only eleven hundred Hawaiian monk seals left in the wild. The risk of losing even one individual in such a tenuous population was too great. The safe course of action was no action, or at least hands-off science, with the hope that nature would eventually gain control.

But I had learned an important lesson after three decades of studying animals: humans had overwhelmed nature's capacity to heal. One by one in my lifetime the large predatory mammals—the lions, bears, whales, tigers, and wolves that I loved—were facing extinction. Monk seals were on the top of the list. I could not bear standing passively by, hoping for a solution to materialize. So I continued to fight a paper battle for this isolated Hawaiian seal.

Now suddenly, out of the Antarctic blue, an e-mail challenged me to step up.

As I gazed out of the frosted window, there was nothing more intriguing for me than going tropical. In one short reply I would shed years of paperwork and three layers of insulated clothing.

I ignored the obvious logistical nightmare of trying to house a tropical Hawaiian seal in the middle of winter in Santa Cruz, California. The bigger problem was that I had no money to care for the animal. Yet I tapped "Y-E-S" on the chilly keyboard, with each letter representing five years of begging for permits.

The odds of an endangered monk seal coming to California were so low as to be laughable. I doubted that I would ever again hear another word about this orphaned seal pup.

The next morning Beau proved me wrong.

3.

First Steps

H e's KP2," Beau announced at breakfast in the McMurdo Station galley.

"KP2?"

"His name. The monk seal is called KP2. It stands for Kauai Pup 2, the second Hawaiian monk seal pup born on Kauai in 2008, and he's a regular celebrity!" Beau had gone into detective mode for most of the frozen night and had found that there was a lot more history to the abandoned seal pup than was relayed in Jennifer's e-mail. Sixteen months had passed since the day of KP2's rescue from his birth beach in Kauai. In that time, the seal had created quite an uproar.

"He's been written up on the front page of the *Wall Street Journal* AND he's been on CBS News. He's even got YouTube videos! There have been protests about him on Molokai and Oahu. This is not just a simple case of a rehabbed seal pup." There was a note of caution in Beau's voice.

I didn't understand why one stranded seal pup had warranted so much attention, even if he was an endangered species. At home in California, hundreds of stranded sea lion, harbor seal, and elephant seal pups were taken into rehabilitation facilities each year. After fattening them up for a couple of months, the facilities scooted them out the door and back to sea with little fanfare. Something was very different about KP2.

Concerned that we were getting in over our heads, Beau had placed another transglobal phone call to Traci in Santa Cruz. Technically Traci was the program manager and training supervisor in my lab, while Beau worked for her as the head trainer. Their positions notwithstanding, they were confidants in all things animal and related as family members rather than just close friends.

Together these two marine mammal trainers were the engineers of science for the animals we worked with every day. I supplied the research ideas, and Beau and Traci translated the tasks into doable behaviors for seals, sea otters, and dolphins. They were a yin and yang animal training team that traveled the world with me. Traci and I shared a tomboy attitude and a childhood shaped by multiple siblings. Growing up in the mountains of Southern California with three sisters, Traci had a childhood that might have shaped her into the ultra feminine. Instead she loved to get dirty and play with power tools. Traci could maneuver a Bobcat forklift with ease and then carry a blue Vera Wang purse to lunch. Research training was her unique skill, her life's work, and her first love.

With Beau I shared a parochial school education and the strict discipline of priests and nuns. While I grew up on the East Coast during the era of the Newark race riots in the sixties, he lived with the pervasive street tension of the eighties on the outskirts of the roughest section of Oakland, California. As a minority white boy in an integrated school, Beau quickly learned to appreciate diversity, a lesson that helped him to overcome the *haole* stigma when he moved to Hawaii. Beau carried those early experiences with him and was dedicated to teaching young children about sharing the planet—both among themselves and with animals.

Beau was passionate; Traci was practical and persistent. It was my good fortune to have such a team on a shared conservation mission. However, both trainers now expressed doubts about taking on a high-profile, orphaned monk seal pup.

I swallowed a spoonful of cold cereal and chided Beau, "Where's your sense of adventure?" I lived by the words that guided M. C. Escher, the Dutch graphic artist whose etchings featured impossible explorations of infinity: *Only those who attempt the absurd can achieve the impossible.*

With the chemical taste of powdered milk dissolving on my tongue (I longed for the luxury of *real* cow's milk on this frozen continent), I thought, what could be more absurd than moving a wild Hawaiian seal born on a tropical island to a university campus in the redwood forests of Northern California? Then again, what if, during our work, we achieved the impossible? What if we were able to help pull an endangered species away from the edge of extinction?

We had to start somewhere.

After breakfast, Beau and I began piecing together the first year and a half of the young monk seal's life from news articles, NMFS reports, and a few phone calls. KP2's first seven months had been spent in the company of humans. Following his momentous flight to

Oahu, the two-day-old pup was placed in the PIFSC Kewalo Research Facility near Honolulu. Repurposed saltwater pools that had once served as a premier research center for studying the biology of large oceanic tuna were his home. To prevent an unforeseen drowning, the pool was drained, leaving the tiny seal pup high and dry on a sterile floor shaded by a small tent.

Instead of lazing in the sun snuggled on soft sand next to his mother, KP2's days were filled with human hands desperately trying to keep him alive. Drs. Gregg Levine, Bob Braun, and Frances Gulland formed an expert veterinary team that would have been the envy of the top dog at Westminster. Their primary cause for concern was the extreme young age of the pup; no one had ever tried to rehabilitate a two-day-old Hawaiian monk seal. Without the immunities provided by his mother's milk, KP2 started life with a severe health handicap. He was a wild baby living in a world of adult human germs.

In pediatrician fashion, the veterinary team compared KP2's size to other island pups. Stretching out a sewing tape measure, the seal's length and girth were measured and recorded on a medical chart. The doctors placed the tiny pup in a plastic Rubbermaid tray mounted on a bathroom-sized scale for his first weigh-in and inserted needles beneath his skin to take blood samples and provide antibiotics.

KP2 endured all the probing and poking without complaint. Despite the intrusiveness of the procedures, it was in these helping hands that the young seal found the nurturing that all young mammals crave. I was impressed to find that more than seventy volunteers from all walks of Hawaiian and Californian life had lined up to help care for the orphaned seal pup with the big, gravelly voice. Their only qualification was simple, unwavering dedication.

The creation of a human family to nurture such a young monk seal was unprecedented. With an abusive mother, an absentee father, and no surviving siblings, KP2 had no immediate seal relatives. Under the best circumstances, there is little that can be construed as "family life" for any species of seal. There is no fatherly interaction. Male seals serve only one purpose: to impregnate females. After sex the prospective fathers disappear, never to be seen by their offspring born approximately a year later. On the maternal side, monk seal mothers are initially attentive and forgo food themselves as they protect and feed their new offspring. But this idyllic maternal period is short-lived. After six weeks, these same mothers abruptly turn off the milk bar and head to sea, turning their backs on youngsters they will never see again.

I'd worked with Weddell seals in Antarctica and harbor seals in California, and found both species to be as independent as cats. Hawaiian monk seals went even one step further in marine mammal aloofness, and tended to be the most solitary of marine seals. This introverted character is one of the reasons behind the name "monk"—that and their dome-headed, spiky-eyebrowed, bewhiskered appearance.

As a result of this behavioral ancestry, the company of seals was not what KP2 craved.

"*Brrauuigh!*" KP2 took little time to inform everyone around him that he was hungry. He had not had a drop to drink or eat since he was born, and his skin sagged on his tiny frame.

In answer to KP2's cries, the seal's veterinary team blended salmon oil, electrolytes, vitamins, and a zoo milk supplement called Multimix into a "salmon shake." The team worked with a sense of urgency, knowing that early nutrition dictated the development of growing tissue in youngsters, particularly the sensitive wiring of the brain, nerves, and eyes. But the seal pup did not know how to eat. A baby bottle held to his lips was as meaningless as the black lava rocks he had suckled to no avail on his birthday.

The veterinarians decided to go for a more direct approach. It was not exactly the maternal response the rumbling pup expected. While one sat astride KP2 and held on to each side of the squirming pup's neck, another snaked a two-foot length of rubber tubing down KP2's esophagus and into his empty stomach. The pup tried to roll and shake the tube out with a growl, but the vets persisted. They attached a large plastic syringe filled with the salmon shake, and with a steady push pumped nutrients into the hungry seal. Slowly the little seal calmed down as his abdomen began to bulge with food. With his stomach full and the taste of salmon burps on his tongue, KP2 fell sound asleep.

Like any two-day-old mammal, KP2 needed to be fed around the clock, and his adoptive volunteers came to the rescue. As the weeks passed, their efforts paid off as KP2 reached many of the proper monk seal milestones. By day twenty-six his first baby teeth (two razor-sharp incisors) had erupted and were cause for celebration. Ten days later he began to molt his black puppy lanugo fur and develop a sleek gray coat that was the uniform of a juvenile monk seal.

With his interminable attraction to humans, KP2 grew both in stature and favor among his caretakers. As he grew older, he followed people like a lost dog trying to con them into staying longer. He bounced around his pool in anticipation of their arrival and his next meal, sliding in a circle along the perimeter with his floppy fore flippers propelling him along. If the caretakers were ever late, he voiced his impatience with a throaty "*brrrooaarr.*"

Before the volunteers' eyes, the pup quickly matured into an independent young seal, and talk of his release back into the wild began to filter among the members of his veterinary team. The thought of KP2 being able to join his fellow monk seals gliding effortlessly through the warm, blue island waters was so exciting for his adoptive family that they were almost afraid to discuss it out loud. They envisioned their little seal sneaking up on *o'opu*, the prickly spotted blowfish, and chasing *honu*, the green sea turtle. Once free, KP2 would feel the song of *koholā*, the humpback whale, reverberate through his body as it had for Hawaiian monk seals for millions of years. All of this was possible except for one problem. KP2 had never entered the water and, unlike any seal his age, he did not know how to swim.

To solve this oversight in his schooling, KP2's pool was immediately filled and swimming lessons began. It was something that his mother should have taught him in the shallow waters of Kauai within weeks of his birth. On his first foray into deep water, KP2 proved that nature knows best. His initial attempts at submerging had his adoptive Hawaiian family rolling with laughter. They threw fish into the water to encourage him to dive, but buoyancy always got the better of the round seal as he eventually rose belly- or tail-first back to the surface. In contrast to the first swim by skinny wild pups, KP2 had grown too fat on salmon shakes to submerge. The thrashing seal bobbed like a diver wearing a thick wetsuit without a weight belt. Swinging his hind flippers frantically from side to side, KP2 churned the water into a white froth. Clearly, this "child of the ocean" needed assistance.

Help arrived in the form of a pink boogie board from his swim coaches. As soon as they tossed the floating board into the pool, KP2 climbed on top. Although he was not much of a diver, the seal quickly proved to onlookers that he was a natural surfer, spending hours paddling around his pool atop the board.

The ubiquitous foam boards would remain a source of comfort and safety for KP2, a life preserver on a journey never before ventured by a Hawaiian seal.

4.

Growth

—

There's something else you need to know about KP2," Beau
reported as I sat at my computer in the marine biology wing
of the Crary Science and Engineering Center at McMurdo
Station. I tried to gauge the gravity of his message but found it difficult
to take him seriously as he dusted snow from the blue flowered board
shorts, bare legs, and hiking boots with no socks that poked from beneath his giant red Antarctic parka. A night of howling wind, sideways
snow, and a bone-chilling drop in temperature had left the expedition members sleepless and bleary-eyed. Joints were stiff and refused to
move; no heater in the dorm buildings could keep up with such blister-
ing cold, so we shivered through the night. I did not even want to
know where Beau had been in his getup.

Beau's second night of Internet surfing after Jennifer's e-mail had uncovered a disturbing series of entries in KP2's early medical history. Initially the signs were easy to ignore. Behaviorally, the little seal was as rambunctious as any rotund Labrador puppy with a pool of his own, splashing, eating, sunbathing, and riding his pink boogie board. However, as the pup's first summer wore on, a change had slowly and subtly occurred.

"KP2 is going blind," Beau murmured as he tried to rub the cold out of his bare calves. My first reaction was to think: what more could happen to this seal, and would his condition affect our proposed science? Why had no one mentioned this to me? But I said nothing and waited for Beau to finish.

"His veterinarians are not sure what caused his eyes to cloud over," he continued. "All of the standard tests for pathogens were negative." Beau took a breath and waited for a reaction from me. With none forthcoming, he finished, "KP2 should have been released back to the wild after being in Kewalo for two months. Instead he was transferred to an ocean pen for another five months of rehab."

As Beau filled me in on the details, I soon had the strange feeling that this little monk seal was following me.

On September 8, 2008, after winding along the lushly vegetated *pali* highway that segregated windward and leeward Oahu, four-month-old KP2 entered the haunts of my old island life. His destination was as far from the unstructured, carefree lifestyle of a wild monk seal as possible; his new home was the U.S. Marine Corps Base Hawaii (MCBH) in Kaneohe.

As a postdoctoral researcher for the U.S. Navy's Dolphin Systems program, I had once lived, played, and worked at the Kaneohe Marine Corps Air Station—the previous name for MCBH. Dolphin Systems were, literally, marine animals. With the same professionalism and sense of duty as the human Navy SEALs, the bottlenose dolphins of the program were trained to detect submerged mines using their incredible internal sonar. I was fascinated by the dolphins' ability to see through objects. Their sonar sense was so exact that they could assess the internal diameter core of a submerged aluminum pipe with more precision than a scientist's micrometer. More astounding was the fact that they were capable of maintaining this accuracy hundreds of yards from the target. I couldn't even see the pipe from that distance, much less guess its inside diameter. Add in the animals' intelligence and you had the perfect "system," as the Navy liked to refer to them, for finding submerged mines and protecting sailors and ships.

Two dolphins I worked with in Kaneohe were Puka, named for a unique hole in his pectoral fin (*puka* means "hole" in Hawaiian), and Primo, named for a brand of Hawaiian beer. My job was to evaluate their swimming and diving capabilities, which meant glorious days out in the bay. These two special dolphins and I began a lifelong friendship in the blue waters of Kaneohe.

Puka and Primo, along with my Welsh Pembroke corgi, Austin, formed my tiny Hawaiian *'ohana*. In Kaneohe I found that I was rich with family, an unusual situation for a wildlife biologist used to a nomadic life of science. In the drive to save endangered wildlife, I had traveled the world and considered the isolation and hardships as part of the adventure. The one sacrifice had been family. In retrospect, I see that singular dedication of my life as very nunlike, although it was not quite the version my parents had envisioned.

Through sheer exuberance, the two gray bottlenose dolphins and Austin showed me the spirit of Hawaiian living. The four of us were inseparable and explored the island of Oahu together with abandon.

Puka and Primo shared the beauty of Oahu's water. On excursions in a Boston Whaler we visited coves and reefs and explored the depths of the windward island. Some days they would race my boat, and on others they would take off, disappearing to depths that I had no hope of reaching even on scuba. Minutes later I'd spy them leaping more than eighteen feet into the air in a display of athleticism that always left me breathless. They were cetacean Olympians that performed for the sheer pleasure of their tightly muscled bodies, and I was thrilled to be in their presence.

With Austin by my side, I explored the lands of Oahu. Surrounded by the intoxicating perfume of plumeria, coconut palms, giant ferns, and tropical forest decay that permeated Kaneohe, we forged new hiking trails up the surrounding *pali*. Unconstrained foliage that threatened to reclaim the roads made the fragrant air cling mosslike to our skin. Living so close to the ground, Austin experienced the island from a whole new perspective. Ubiquitous strangler fig roots weaving across hiking trails were as monstrous as fallen trees to the short dog, and heat rising from black volcanic surfaces created a sweltering microclimate in his thick, dark coat. On those occasions he had to be carried on my shoulders, where the windward breezes cooled his batlike ears. I loved that he was oblivious to his handicapped stature and that I could help. We were 'ohana.

I imagined KP2 traveling across the island, sniffing like Austin had at the passing banana plantations and the two-story wood-framed houses outside the guarded entrance to the military base. Along the way he must have passed the Koa House, where the best banana pancakes in the world sizzled daily on a griddle and were served with a side of warm coconut syrup.

With the assistance of the MCBH Environmental Division staff, the orphaned, nearly blind seal pup was given a private cove normally reserved for visiting dignitaries. Lava rocks, mangroves, and plastic netting submerged and stretched across the entrance of the cove created a secure ocean pen for KP2. I could not think of a better place to help the little seal recuperate mentally and physically from the trauma of his first months of life.

Once released from his kennel, the pup splashed and rolled in the shallow waves that swept onto the sand. An entire summer had passed since he'd last burrowed on a warm sandy beach. In the frantic days following his birth, he'd never had the opportunity to listen to the surf or smell the salt-laced sea air. But in Kaneohe, KP2 could crawl among the volcanic rocks strewn along the coastline and explore the warm tide pools.

Like most young boys introduced to water, KP2 immediately invented spitting and splashing games. As his caretakers looked on, the seal gulped seawater and then spewed it upward like a fountain over and over again. When they tossed in a green coconut that had fallen from one of the surrounding palm trees, KP2 learned about buoyancy. He would hold the coconut under the water with his paddle-like front flippers and then let go. Each time the coconut blasted to the surface and plunked down with a resounding splash, only to have the seal throw his body on it to begin the game again.

When the tide rose, KP2 finally had a chance to really test his swimming muscles. Tapered on both ends, KP2's body, like that of all marine mammals, was perfection in streamlining. His nose-to-tail length was four times the maximum diameter of his body, determined by physicists to be the ideal proportions for reducing hydrodynamic drag. Anatomical features that could have created turbulence— limbs, testicles, and penis—were tucked inside. A thick blubbery layer smoothed his contours. KP2 slipped through the water quickly and stealthily, and he suddenly was cognizant of his unique advantage over fish as well as human swimmers.

News of the orphaned seal pup traveled quickly across the base and intrigued the military residents. During the night, military police patrolled the cove to offer protection for the seal pup and his volunteer caretakers. The little Hawaiian monk seal that had been abandoned by his own species was now being safeguarded by the U.S. Marine Corps' finest. An enforcement officer began delivering eight-pound live *tako* (octopus), which quickly turned into a favorite play item for the seal. KP2 tossed the octopus with eight legs swinging through the air and then hauled it in his mouth back and forth across his pen. When he finally determined that he had killed his prey, he swallowed it. In the secluded cove, KP2 began returning to his species's roots and developed into the ocean predator he was destined biologically to become.

In the freedom of his net pen, surrounded by other ocean creatures, KP2 also revealed an unexpected and amazing talent. Under the watchful eyes of his caretakers, he displayed a behavior never before observed for a phocid seal.

Butting his head along the sandy bottom, KP2 began pushing small rocks and broken coral bits into the deep corner of his enclosure. He carefully piled the submerged rocks atop one another. Then he waited. Before long, small fish, crabs, and sea cucumbers were attracted to the dark crevices of his artificial reef. At first the fish and invertebrates were a source of entertainment for the curious seal. Then they were dinner.

"Oh my gosh, Beau!" I exclaimed when I heard about the seal's early hunting behaviors. "KP2 created a tool!"

Like many tools developed by animals, including man, the young seal used his for feeding. In the thirty years that I'd been studying seals, I had never encountered one that made or used a tool. Weddell seals will blow bubbles into brash ice to flush out fish, but that is not the same thing as taking an inanimate object and forming it into an article to serve a unique purpose. That is the definition of a tool, and few animal species are capable of mastering this feat. Such discoveries are so rare that they are generally heralded by the public and scientific community as unique evidence for animal intelligence.

Clever gorillas, chimpanzees, and orangutans are famous for fashioning sticks into spears, clubs, and even ant picks. Elephants swat flies with branches held in their trunks, and innovative bottlenose dolphins in Western Australia wear bits of marine sponge to protect their rostrums as they dig for fish buried in coastal sediments. Sea otters and sea gulls use rocks as hammers to open shellfish.

Now among this distinguished group of intelligent tool-using animals was KP2.

LIVING AT THE FROZEN BOTTOM of the world, three thousand miles from the Hawaiian Islands, I recognized once again that there was something very different about KP2, both in how he interacted with his environment and in how he interacted with people. It was almost as if he could read other creatures. Consequently, he affected the people he met, instilling them with a joie de vivre. Everyone, from veterinarians to the U.S. Coast Guard and U.S. Marine Corps to an army of volunteer caretakers, had fallen under his spell. They readily sacrificed their time to ensure his survival, even if it required teaching this naive seal how to eat and how to swim.

In contrast to the independent, catlike Weddell seals I was studying, which weathered Antarctic extremes with stoic resolve, KP2 exuded optimism in the face of hardship. Despite all that had happened to him in the first months of his life, KP2 had remained boisterous, curious, and spirited.

With this realization, I vowed to be careful around this animal. I was a scientist first and resolved that I would not fall into KP2's emotional trap. There were no Kleenex boxes in my office at the university; there was no room for crying in science. No cute animal names ever made the pages of my publications. There was no anthropomorphizing. Emotions were a female luxury sacrificed long ago in order to survive in the male-dominated world of science. I would not change now.

KP2 was a scientific opportunity—nothing more, nothing less— and I knew that I needed to remain objective if his science was to be taken seriously. The seal's personality could not dominate our mission. Hugs and tears never saved an endangered species.

Still, I was curious: if KP2 was so special to the people of Hawaii and so important for his species, why did he have to leave the islands?

5.

Discovery

—

After constructing his feeding reef, KP2 began to move toward complete dietary independence. Instead of relying solely on his human companions for sustenance, KP2 started to supplement his diet with natural items from the sea that he caught himself. He still enthusiastically accepted *tako* handouts, but more as a toy than a necessary part of his food intake.

Over the months as KP2 matured and swam for progressively longer periods, the ocean water slowly improved his eyes. By Thanksgiving the corneal swelling and cloudiness had regressed. Soon thereafter discussions about the young seal's release back into the wild resumed in earnest.

Encouraged by his progress, KP2's vets moved quickly. Thus, two weeks later, as his first Christmas approached, KP2 was given the best present that could be imagined for a wild Hawaiian monk seal: the island of Molokai.

AFTER PASSING A RIGOROUS series of veterinary examinations, KP2 was bundled once again into a cage and flown over the blue tropical interisland waters. His eyesight was not perfect, but he had proven that he could see well enough to catch fish. In fact, he had been so successful that the U.S. Coast Guard crew had trouble lifting his cage into their HH-65 helicopter for the transport. KP2 had gained 115 pounds, and now weighed more than a full-grown Great Dane.

His caretakers wondered how KP2 would view his freedom. The young seal had lived for 227 days in the company of humans, five months longer than normal rehabilitation programs for newborn seals and sea lions. Seal pups his age were already living independently. They would soon discover that while KP2 owed his life to their dedicated hands, this long-term contact with man would prove to be both blessing and curse as he reentered the wild waters of Hawaii.

KP2'S NEWEST DESTINATION was one of the most remote areas of the main Hawaiian Islands, the Kalaupapa Peninsula of Molokai. With Molokai shaped like a dolphin swimming eastward, the peninsula forms a five-square-mile flat plate of land located at the dorsal fin. White churning waters that crash onto the shore and soaring *nā pali* (sea cliffs) sixteen hundred feet high isolate the peninsula from the rest of the island. Since the first volcanic explosions that gave birth to the island, this impenetrable jut of land has served as both paradise and prison.

The young seal's transfer would be the second time in Molokai's history that the U.S. government had taken advantage of the geographical isolation of the area. In each case the goal was to prevent human contact. For KP2, the isolation was meant to wean him from dependency on humans. For the souls who had walked the sands of Kalaupapa Peninsula nearly a century and a half before, the isolated peninsula was a natural quarantine facility.

In the late 1800s, in a desperate attempt to quell the Hawaiian epidemic of Hansen's disease (formerly called leprosy), the U.S. government shipped more than eight thousand patients to Kalaupapa Peninsula. The chronic, disfiguring disease robbed victims of their eyesight and the use of their hands and feet. The government program robbed them of families and friends. Segregation of the victims lasted for more than one hundred years, until the availability of sulfone-based drugs put the disease in remission. With Hansen's patients no longer contagious, the isolation laws of Hawaii were abolished in 1969, leaving a remnant settlement and a proud, close-knit, and exceedingly friendly community of nearly 150 patients and workers.

In recent years, the residents welcomed the arrival of a small cohort of female Hawaiian monk seals attracted to the isolated beaches of Kalaupapa Peninsula. Sharing the sands with the community, the seal mothers gave birth and raised their pups in the solitude of the natural nursery.

It was into this intimate human and seal community that the National Marine Fisheries Service, with the help of the U.S. Coast Guard, delivered an energetic, cloudy-eyed KP2 ten days before Christmas. He quickly proved that his disability did not mean that he was immobile.

NMFS SCIENTISTS ATTACHED two small tags with epoxy to the fur on KP2's back to track his movements, much like the ankle monitor worn by individuals placed under house arrest. Using the same GPS technology that allows an iPhone to know where you are calling from, the scientists were able to follow KP2's exploits.

According to David Schofield, the NMFS-PIRO stranding coordinator who had originally rescued KP2 on Kauai, "The release site was selected primarily because it is used regularly by other seals and would provide the opportunity for KP2 to socialize." The tags would show that KP2 clearly had other ideas.

For the first month the young seal seemed reluctant to venture from his release site. Staying close to Kalaupapa, KP2 briefly joined up with another young male monk seal to play-fight in the shallow waters. NMFS scientists were encouraged.

"All signs point towards KP2 successfully reintegrating into Hawaiian monk seal society," David reported to the Hawaiian Monk Seal Recovery Team in December.

Things changed rapidly after that. By January, KP2 had grown bolder and began cruising the rugged northern shoreline of Molokai, where the world's largest sea cliffs rise dramatically from the water to the sky. Passing towering two-thousand-foot waterfalls, the seal crossed the Kaiwi Channel on a track back to Honolulu. Then he abruptly changed his course and returned to Kalaupapa Peninsula, completing a forty-mile trek that belied his young age.

By February, KP2 was routinely circumnavigating Molokai and diving for fish in the Kaiwi Channel. On a landmark day that month, the adventurous seal swam south to the island of Lanai, where his natural curiosity drew him to the fishing activities of Kaumalapau Harbor. He stayed only briefly on Lanai, and was lured back to the "dolphin belly" of Molokai. The same curiosity and familiarity with humans brought the young seal to a small Hawaiian town with the tongue-twisting name of Kaunakakai.

The 450-foot wharf of Kaunakakai was alive with paddlers, kayakers, and recreational and commercial boats. On the opposite side of the wharf the splashing of children caught KP2's attention. Having never seen such young swimmers, he cruised over to the laughing youngsters with a mixture of suspicion and inquisitiveness. He circled and dove around them, popping his head up for an occasional breath that elicited squeals of delight. Before long, word about the "friendly seal down at the wharf" had brought others from town.

What happened next began as a simple decision all too familiar to any swimmer in Hawaiian waters. Yet this one act triggered a life-altering event for KP2 that would eventually instigate an international chain reaction, from the islands to Washington, D.C., to Antarctica and back.

Hearing about KP2's antics, Alona Demmers had rushed down to Kaunakakai Wharf with her children. She eagerly jumped into the water, hoping to get a closer look at the cute monk seal. In her hands she held a float for her kids to rest on. Unknowingly, Alona had brought the one thing that KP2 would find irresistible: a pink boogie board.

The seal immediately recognized his favorite toy from Kewalo Basin and knew exactly what to do. He hopped on top and began to paddle. From that moment on, it was all play for the seal and the children of Molokai.

There was one child in particular KP2 actively sought out each day. "The dog that ran in rough waters," as his Hawaiian name described him, appeared to bond in canine fashion with a young boy with large, dark eyes and an endless grin. Kalaekahi, nicknamed Kahi, was an energetic eleven-year-old Hawaiian who appealed to KP2's love of roughhousing. With similar stocky builds, the two bear-hugged each other and rolled in the water like rowdy puppies. The boy became inseparable from the seal, and KP2 followed him with doglike zeal.

. . .

Officials from the National Marine Fisheries Service in Hawaii, operating under NOAA, its parent organization in Washington, D.C., had specifically released KP2 at Kalaupapa Peninsula so he could learn how to be a seal. Instead, each day he was becoming more and more human as the friendly seal charmed the locals with his surfing and played water games with anyone willing to jump in the water with him. Neither state nor mainland officials were amused by KP2's summertime antics.

To stop these potentially dangerous interactions, NOAA sent representatives to the island to urge the locals to avoid playing with KP2. But ignoring the charismatic seal was impossible. For some, KP2 was a mascot, a creature from the sea who had chosen to live among the people of Molokai. KP2's story, from his abandonment to his rehabilitation and release, appealed to them; there were so many parallels with the island's history. Ignoring such a kindred spirit was particularly difficult now that the seal frequently snoozed on the wharf, the boat ramp, and even on the stern dive platforms of the boats tied up to the dock. People were literally tripping over the sleeping seal.

However, one island group was in agreement with NOAA and equally upset by KP2's presence in the harbor. To many local fishermen, KP2 represented a voracious predator that would eat up the island's fish and get entangled in their gear. Where there was one hungry seal, there were surely more to follow. They demanded that NOAA remove the fish-stealing seal or they would take matters into their own hands.

Hearing rumors around town, Alona Demmers became concerned for KP2's safety. One day she watched in horror as a fisherman kicked the sleeping seal several times to move him out of the way. But KP2 did not react. The kick was just another human behavior. Humans had poked him with needles, put him in cages, and given him medicines. They had also fed him, played with him, and were a source of immense comfort. During his short life, KP2 had already learned that the nature of man was unpredictable. So he merely rolled over and continued to sleep.

IN MARCH AN NMFS TEAM dispatched from Oahu arrived on Molokai and herded the sleeping seal into a cage. He was weighed, cage and all, with the boat davit on Kaunakakai Wharf. Hanging beneath the scale like a prized marlin, KP2 weighed in at just 98.5 pounds. He had lost nearly 34 percent of his body weight since his release, causing some to question his dietary habits while playing around the wharf. After checking his overall health, the team transported the playful seal several miles up the coast to discourage his behavioral interactions with the children.

With perfect navigation, KP2 defied the NMFS's plan and immediately swam back to Kaunakakai and his waiting companions. NMFS volunteers tried shaking palm fronds to scare the seal away from the wharf. Although initially spooked, the persistent young seal returned again and again. Finally, his dedication won out. The palm frond shakers gave up and left KP2 to play with Kahi and the rest of the splashing, laughing swimmers for the rest of the summer.

FRICTION IN THE COMMUNITY GREW as the seal frolicked with his young friends. The children of Kaunakakai loved the one-year-old seal, while fishermen trying to scrape out a living from a once endless ocean bounty grew increasingly disgruntled. NOAA and NMFS officials found themselves stuck between a mascot and a nuisance of their creation.

The breaking point came on June 10, when a resident of Molokai posted a 1:21 minute video entitled "Mac and KP2" on YouTube. In it KP2 chases a confused yellow Labrador retriever out of the water by the wharf. The streamlined seal has the obvious advantage as he flipper-slaps the water with a resounding splash right in front of Mac's fuzzy muzzle. Peals of laughter can be heard from onlookers in the background as the dog hightails it for land. To the dismay of NOAA, the "friendly seal of Molokai" suddenly developed an Internet fan club.

There was much more behind NOAA's concern than just friendly people. Seeing the seal and dog interact raised a concern that went beyond the safety of KP2 and the children of Molokai. This time the entire fragile Hawaiian monk seal population was at risk. Dogs and wild seals were not supposed to mix; their genetics were so closely related that diseases could easily pass between them. With no natural immunity to diseases such as distemper or rabies and without the benefit of inoculations, wild monk seals, including KP2, were highly vulnerable to terrestrial-borne diseases. Even if KP2 did not get sick, he could carry the viruses to other immuno-naive monk seals, to the demise of his entire fragile species.

Such a viral epidemic among seals had occurred in 1988 and again in 2002, when an outbreak of distemper resulted in catastrophic losses of European harbor seals. More than twenty-three thousand harbor seals died in the 1988 event; another thirty thousand succumbed to the disease in 2002. Thousands of seal bodies littered shorelines across Denmark, Sweden, Germany, and the Netherlands, and then in Britain and Ireland. The eleven hundred remaining Hawaiian monk seals would never survive such a devastating epidemic.

With the YouTube images fresh on their minds, NOAA and NMFS officials once again corralled KP2 into his cage. This time KP2 was herded back to his original release point on Kalaupapa Peninsula, more than forty miles away on the "dolphin dorsal fin" side of Molokai.

But KP2 knew the game. It took him less than forty-eight hours to navigate around the island of Molokai and arrive back at Kaunakakai Wharf. In his previous travels he had mapped the currents and water channels of the island in his head. As a result, by sunset on the second day following his transfer, the seal was back playing with the swimming children, who waited in the glow of approaching dusk for his return.

Thwarted by the persistence of KP2, government officials pleaded with the locals to leave the seal alone. Unfortunately, his celebrity spread nationally when Chris Herring wrote an article about KP2 for the *Wall Street Journal*. From New York City to Honolulu, KP2's front-page story and pictures made people smile.

The tale of the boisterous baby seal that loved people immediately fed into the national media. That evening CBS News carried a news report on the "friendly seal pup that acts like a puppy." David Schofield from the NMFS Honolulu office tried an on-camera warning about swimming with KP2 for the broadcast. But the message was unheard in KP2's wake. Even the journalists cracked a smile at KP2 as he enthusiastically sidled up to grinning swimmers, flopped on boogie boards, chummed it up with paddlers, and played with the children of Molokai. The message that the public heard was not one of danger, it was of fun.

WHILE BEAU AND I were shivering on the ice three thousand miles away, KP2 made an end-of-summer error that would change all of our lives. On a quiet September morning the seventeen-month-old seal swam up behind an unsuspecting swimmer near Kaunakakai Wharf and bear-hugged the person in his usual rambunctious greeting. This time, instead of a laughing child, the swimmer turned out to be Ingrid Toth, a thin, white-haired seventy-year-old woman. Despite his young age, the seal outweighed Ingrid and was able to hold her underwater for several seconds. The justifiably frightened woman suddenly had second thoughts about swimming near the wharf, and NOAA officials had finally heard enough.

Plans were immediately made to permanently move KP2 away from Molokai. But where? He was neither wild nor tame. His months in rehabilitation with humans and subsequent release among seals had shaped him into a creature that had no home.

Instead of fearing or attacking humans, KP2 had chosen to live among them. And that made KP2 both irresistible and dangerous, depending on whom one wanted to believe.

Jennifer Skidmore of the Office of Protected Resources began exploring the options for KP2 from behind an NMFS desk in metro Washington, D.C. Two Hawaiian aquariums, Sea Life Park and the Waikiki Aquarium, declined to take the young monk seal, citing lack of space. No other facilities for holding monk seals existed in the state.

Under pressure from an increasingly volatile community in Molokai, Jennifer broadened her search to the remote ends of the earth by sending an e-mail to my Antarctic camp on ice. "We have a situation with a Hawaiian monk seal pup . . ."

IN LESS TIME than it took for Jennifer's e-mail to reach me near the South Pole, the people of Molokai heard about KP2's pending removal by NOAA. The community exploded. The owner of a local restaurant, Darrin Abell, told the *Wall Street Journal*, "If they ship KP2 off the island, it might get ugly here. There'd be an uproar." Similar sentiments were expressed by Walter Ritte, a longtime activist for Hawaiian culture who was especially vocal about the importance of KP2 to the people of Molokai. "If they take him away one more time and he comes back," he told the reporter, "I don't think this community will let NOAA take him again." Even Ingrid Toth, the frightened swimmer, stood behind KP2's choice of Kaunakakai Wharf as his native home.

On the opposite side of a stuffy room where people discussed KP2's fate during a town hall meeting, fishermen voiced the opposing opinion. They wanted KP2 gone. They knew of older fishermen who had shot monk seals to protect their livelihood. This seal, with or without a name, with or without a media following, would be treated no differently.

WHILE THE MAELSTROM brewed on Molokai, options for KP2 quickly evaporated.

Resistance to the unprecedented move of the Hawaiian seal pup to my mainland lab was swift. Members of the Hawaiian Monk Seal Recovery Team immediately rejected the idea of moving a member of an endangered species. No one had ever transferred a young Hawaiian monk seal out of the islands. The Marine Mammal Commission, an independent U.S. agency that provides oversight of marine mammal conservation policies, also expressed reservations. Veterinarians, scientists, Washington administrators, personnel in the NMFS permit office, and marine mammal curators from Sea Life Park to SeaWorld debated. Most of them were part of the small community of marine mammal science and knew me personally. They knew the quality of my research. But once again the arguments had little to do with science; rather, KP2's fate hinged on politics and precedent.

After several weeks Amy Sloan, my permit officer from the same office at OPR as Jennifer Skidmore, finally informed me, "You've got enemies as well as supporters." I trusted Amy. She was careful enough not to provide details, saying only that she and Jennifer were fighting an uphill battle back in Washington.

Amy knew both sides of the battlefield when it came to marine mammal research. Before landing behind a desk in the OPR permit office, she had worked in Hawaii on the NMFS Hawaiian monk seal project. She was a self-proclaimed "pinnipedophile," preferring to study the independent seals and sea lions rather than the charismatic dolphins and whales that the public seemed to dote on. "Too rubbery," as Amy had once summarized her feelings about dolphins to me.

Arguments crisscrossing the oceans continued until one communication stopped me cold.

"There's talk of euthanasia," one of the veterinarians leaked to me one day. A rumor had spread from Washington.

"You mean people back there are saying they would rather *kill* KP2 instead of allowing us to transfer an endangered animal to my lab for *science*?!" I was astonished and then furious. The rumor confirmed everything I believed about D.C. bureaucrats and the hopeless situation for wild animals.

Amy tried to explain the logic. "KP2 is a male. He's expendable." But I knew that even she was not convinced by her own words.

How differently we treated the sexes in human and animal worlds. In the world of animal preservation, each female contributed to future generations by creating and nurturing the young. Conversely, males were secondary. One male could impregnate many females, making individual male animals comparatively less important when it came to the reproductive viability of a population and species survival.

For Hawaiian monk seals, males were relegated to an even lower status. With the continued decline in seal numbers during the past thirty years, the sex ratio of the population had become skewed in favor of males. In some areas males outnumbered females more than two to one. Since adult females give birth to only one pup each year, the population needed more healthy females if it was to grow. Thus, from the perspective of the Hawaiian monk seal population, KP2 was another expendable male.

Combined with all the problems he had unwittingly created, from potential disease transmission to scaring swimmers to polarizing the Molokai community, there was little reason to save KP2. Everything could be solved easily with a needle full of Euthasol.

"Don't do or say anything," Amy Sloan warned on a conference call to McMurdo. "We'll deal with it from this end."

Jennifer quickly changed the subject. "How do you transfer your animals?"

"FedEx," I responded mechanically. "Ever since we flew sick sea otters during the Exxon *Valdez* oil spill we've used FedEx. They've always been game for any wet animal we had to transport." I didn't see the point of the discussion if KP2 was going to be killed.

Within a few days Jennifer e-mailed back, "It's going to cost $17,000 to FedEx KP2. Can you cover that?"

"Of course!" I replied without hesitation or thinking. It was the right thing to do.

After my committing to the transport expenses, all communication from Washington suddenly fell silent, leaving me isolated on the Antarctic ice. Days passed and then weeks as one snowy day joined seamlessly with the next.

Our winter research expedition with the Weddell seals ended with the permanent rising of the sun in mid-October. There would be no more sunsets, no more sunrises in Antarctica for the remainder of 2009. Instead the sun would circle monotonously around the horizon. It was time for my team to go home.

ON THE ICE with the cold trickling into the soles of my Bunny boots and the sun circling dizzily around my head, I had a sickening thought on the eve of heading back north.

Where was I going to find $17,000 to save a homeless monk seal pup from Hawaii?

6.

Stolen Child

—

Peering across the edge of the fiberglass pool at the Waikiki Aquarium, KP2 and I regarded each other with a sense of anticipation. I silently tried to weigh the pros and cons of helping this orphaned seal by reading his eyes; he used the same technique to determine whether I was going to hand him something to eat.

"You're not so big for someone who is causing so much trouble," I murmured as I moved away.

Suddenly, I heard a low thumping behind me and reflexively turned around to look out toward the street. The sound was almost inaudible, and then rapidly gained in intensity. It made the hairs on the back of my neck stand up. The rumbling grew steadily and ended in a burst out *"brrawwrrrr!"*

I swung back around to the pool and discovered that KP2 was the source of the racket.

"*Rraaaughhrr.*" The seal pup growled in a low, gravelly rumble that reminded me of Marlon Brando. He then splashed in circles around the shallow water of his pool.

KP2's vocalizations were as adept as any baby's cries in manipulating those around him. In the pool next to his, an adult male monk seal named Nuka'au ("sleek swimmer") also swam in circles; his agitation grew and he growled threateningly back at the pup since he could hear but not see the competitor male who had just entered his territory. KP2's caretakers soon came running with buckets of fish in response to his calls, and to quell the anxiety of neighboring Nuka'au.

In that moment, based on little more than a rumbling growl and a gut reaction to a first encounter, I made the final decision. Somehow I would bring KP2 to California. I also swore once more that I would not become one of his slavish, adoring fans as I watched his caretakers hover and fawn over him. Instead KP2 would be a test, a proof of concept. The seal, together with Traci, Beau, and me, would prove to everyone back in Washington that important science could be conducted on a captive member of an endangered species and used to help the wild population.

I had entered the halls of academia and the field of animal physiology to learn what made each species unique, to understand their limitations and capacities, to see the world through wild animal eyes. I called it Survival Physiology, the identification of each animal's biological Achilles' heel. From that knowledge base, I could design conservation plans based on the biological needs of each species rather than on the leftovers from humans. I wanted to do this for the Hawaiian monk seal before it was too late. KP2 would be the first of his kind to participate in such an experiment. He was a scientific opportunity—nothing more, nothing less.

I did, however, empathize with the little seal's irritability—the previous week had been a blur for both of us. Independently, KP2 and I had experienced radical changes in our environments that eventually dropped us together in Waikiki for our first rumbling encounter.

NOAA had instigated the seal's latest journey. Once the government agency decided to remove KP2 from Molokai, it acted without hesitation. On October 15, KP2 was awakened on Kaunakakai Wharf and crowded into a cage. By the next morning, the journeyman seal was on another U.S. Coast Guard plane, this time headed to the air station at Barbers Point on Oahu. By lunchtime he was swimming in a shallow pool in the back holding area of the Waikiki Aquarium. For the second time in his life, KP2's habitat had shrunk from the expansive luxury of tropical paradise to the sterility of a blue fiberglass-sided pool. He did not know where he was.

While KP2 was learning how to adapt to civilization after months of freedom in the wild, Beau and I were undergoing a similar psychological transition as we finally came in from the ice. For three months we had lived in cold, dark isolation. It took five flying hours from Antarctica to New Zealand, four to Sydney, and ten to Hawaii, dragging us across time zones, the Antarctic Circle, the equator, and the international date line to finally land in Honolulu.

In the city, the cross talk of multiple conversations on cell phones, traffic noise, and the sheer speed at which people moved and spoke overwhelmed us. The sensory overload left me anxious. Now I understood why wild animals in captivity often stare into the distance; I found myself looking past walls, trying to process the confusing haze of activity and sounds around me. I had a whole new appreciation for how wild creatures must feel when suddenly placed in the human world. What I found peculiar was KP2's attraction to it. While I had spent a lifetime trying to enter the animals' world, he had been trying to do the same thing in the opposite direction, and enter mine.

On October 21, five days after KP2's arrival on Oahu, Beau and I landed at the Honolulu International Airport too excited about the scent of green vegetation, the feel of moisture on our skin, and the warmth of the air to care about the jet lag. It was a particularly hot fall day and I realized that the air temperature was over 150 degrees warmer than we'd been used to in Antarctica. As I sweated through my shirt, I marveled at the ability of the human body to endure such extremes.

Beau and I were ignorant of the big events occurring across town with KP2 as we collected our expedition gear from U.S. Customs. Walter Ritte, the activist from Molokai, had flown in earlier with several island residents to stage a protest against KP2's removal from their island. The demonstration in front of the Waikiki Aquarium attracted the local Honolulu news station and provided an opportunity for Walter to vent his views about the treatment of Hawaiians in general, and the Hawaiian seal in particular, at the hands of the federal government. Seeing the front page of the *Honolulu Advertiser*, I wondered who was this wiry Molokai local with island-weathered skin and salt-and-pepper hair. A photograph showed Walter conducting interviews as a small group from Kaunakakai held placards reading NOAA LIED, BRING OUR SEAL HOME, and WE LOVE KP2.

Later I found out that the loss of the friendly seal was upsetting enough, but it was the suddenness of his departure that had precipitated the protest. Children had awakened to find their swimming companion missing. Their parents were furious and so was Walter.

"The kids loved that seal," one Molokai resident told reporters. "There was no opportunity to say good-bye. Nothing."

Walter demanded the return of KP2 to Molokai, where the seal could live in an abandoned fishpond.

Once again David Schofield faced the media spotlight and public mistrust as the federal official trying to explain the reasoning behind NOAA's urgent actions. In a neatly pressed Hawaiian shirt and horn-rimmed glasses, David remained cool in the face of the protesters, which infuriated some in the crowd. He tried logic, recounting the many warnings that had been issued over the past seven months about the consequences of continued human interaction with KP2, but to no avail.

Walter and the people from Molokai remained adamant. The friction between Hawaiian locals and the government was all too evident on camera. The Hawaiians considered the little seal a special gift from the ocean, a *ho'ailona* who had chosen them, not the other way around. To NOAA, humans had encouraged the seal to behave more like a pet than a wild animal.

Throughout the arguments there remained one critical question that neither side could answer. What would happen to KP2's human-friendly ways when he eventually turned into a four-hundred-pound sexually active adult male seal?

By the next morning, the protestors had cleared from the entrance of the aquarium, leaving me alone with the seal and his caretakers. While KP2's greeting was rumbling and enthusiastic, the reception by his human entourage was considerably more guarded. Soon I was sweating more from the scrutiny of KP2's caretakers than from the triple-digit increase in air temperature.

I decided not to reveal my decision regarding KP2's move during the meeting with David Schofield, members from the Hawaiian Monk Seal Recovery Team, the seal's caretakers, and the remaining aquarium staff. Instead I did as I would when encountering any group of wild animals: I simply listened. In doing so I learned one important fact. A veterinary exam upon KP2's arrival at the aquarium had revealed that the seal's eyesight had deteriorated significantly during his escapades in Molokai. Cataracts now obscured 80 percent of his vision.

The news made me wonder. If a male Hawaiian monk seal was considered "expendable" by my opponents, what would a *blind* male Hawaiian monk seal be worth? This could be the final blow for the orphaned seal splashing in front of me at the bottom of his pool.

As I walked out of the aquarium and past the protest area into a nearby park, I was immediately confronted by a pair of women I recognized as volunteer caretakers for the monk seal.

"So what do you think of our KP2?" The two women looked at me with a mix of suspicion and curiosity, circling like Nuka'au in the pool next to KP2. I had no way of knowing what these women had been told or what they thought of my arrival on their island. I got the feeling that it was not all good.

Unsure how to respond, I considered their situation. They had raised KP2 from his very first days, nurturing him in the hopes of his successful return to the wild. Not only was that dream in jeopardy, but now there was a chance that an unknown scientist who'd just breezed in from the South Pole would take him away from the islands forever. I knew what they really wanted to ask: Are you going to do mean scientific things to our KP2?

"I think you've done an amazing job," I finally told them guardedly. "I'd like to help make that effort worthwhile for his species."

There was not a lot more to add. Despite my personal decision regarding KP2's future, I knew that the chances of the monk seal being transferred to my lab were so low as to be laughable. There were so many obstacles, not the least of which was the $17,000 transoceanic airfare.

After a moment I offered, "Maybe there is something that could be done for the people in Molokai. Perhaps the children could name the seal?"

"Oh, they've named him already," one of the women quickly replied. She was wearing a ball cap with a picture of a monk seal sewn onto it.

"Yes," her friend chimed in. "He already has a Hawaiian name." She then quickly rattled off a long series of hard consonants and vowels that were immediately lost on my *haole* ears.

"Oh, good," I said, encouraged that others had shared my idea of a Hawaiian name for KP2. "What does it mean?"

"Stolen child."

My spirits sank as we parted.

CRESTFALLEN, I HEADED DOWN to the water's edge feeling every bit the Dr. Kidnapper. As I dug my toes into the warm, soft sand of Waikiki Beach, I tried to understand how once again my good intentions on the part of animals led to a direct conflict with people. Life would be so much easier if the animals of the oceans could survive just by being left alone. But the world didn't work that way. Man's influence was far too great, and species continued to disappear before our eyes. Something had to change, and in my mind the only reasonable tool was science. Data and the truth it revealed had the power to alter people's thinking.

Early on in my career, I'd recognized that I would have to earn a doctorate degree in science to be taken seriously. I attended Rutgers University in the heart of urban New Jersey in the turbulent seventies. Campuses were still reeling from the Newark race riots, and the main university campus had just voted to allow women to attend classes. I weathered a withering assessment by the blustery, old-school graduate department chair as he tried to convince my PhD adviser that I was not worth the investment since "she will only go off and get married." Despite the dawn of the women's liberation movement, sexual harassment and innuendos were a constant in the classroom, in the lab, and in the field. Glass ceilings, inequality of pay, and discrimination would catch me off guard, but they never dissuaded me. That was the price of gaining knowledge at that time, and I had the satisfaction of doing exactly what I knew I had been born to do. That was good enough.

In retrospect, my difficulties in school prepared me for what was to come as a scientist who worked with big charismatic animals. No sooner had I finished my schooling than the animal rights movement began. Suddenly, my entire chosen field of study was vilified. In 1980, People for the Ethical Treatment of Animals (PETA) was founded. Their motto—"Animals are not ours to eat, wear, experiment on, or use for entertainment"—placed all biologists under intense public scrutiny. The Humane Society of the United States, the Animal Liberation Front, and many other animal protection groups flourished with public donations, and "research" became a dirty word overnight. Scientists, and more specifically biologists, began to rank somewhere below dirt in the public's mind. Unlike veterinarians, we were viewed as users, not saviors, of animals.

"STOLEN CHILD," I REPEATED to myself as the sun began to set. Did I really need another group angry with me? What if the people of

Molokai appealed to Pele, the Hawaiian goddess of fire, to rain bad luck on me as she did to tourists who removed volcanic rocks from her islands? Surely the punishment for absconding with one of her monk seal pups would be even worse.

Staring into the receding sunset, I recalled the moment that I finished my PhD in environmental and exercise physiology and presented the thesis to my father. Disappointed that I was not going to enter the convent and that I'd switched from medical school to become a "sorta doctor," he had asked me incredulously, "You really like animals, don't you?"

"They're my life," I'd told him, and from that point forward, every working day and playing hour had been in the company of animals. Among wild creatures I found a freedom, thrill, and inspiration that has never disappointed. Over the years, I'd been fortunate enough to breathe the dust kicked up from cheetahs sprinting across the African savanna, and feel the pounding heart of an elephant as I rested my head against its massive chest. Playful dolphins had surfed in the wake of my research boats, cutting through salt water with ease, while sea otters had made me laugh with their mittened paws raised in surrender as we passed by. I'd lived among the polar birds, seals, and whales, sharing the explosive crash of calving icebergs all the while marveling at their ability to avoid frostbite on exposed flesh. I was in awe of animals, and indebted to them for the joy they had brought me.

I would never abandon them. Nor would I abandon KP2's monk seal family.

PART II

Passages

Clockwise from left: Traci Kendall, Beau Richter,
Terrie Williams, and KP2

7.

Journeys

KP2's longest journey began with a prayer. On November 18, one week before his transport, several members of the Molokai Island community returned to Oahu and the site of their protest at the Waikiki Aquarium. This time the visit was about safe passages.

In a private ceremony, the orphaned seal received a blessing. In attendance were representatives from every group that had played a role in KP2's life, including his caretakers, the protesters from Molokai, David Schofield with members of the NOAA and NMFS teams, and Beau Richter from my lab, wearing his most formal board shorts. Everyone gathered around the seal's temporary holding pool in the back lot of the aquarium.

As KP2 looked on from the water, Kahu David Ka'upu, a retired

Kamehameha Schools chaplain, and Molokai kumu hula Kanoe Davis performed a traditional Hawaiian blessing. Chanting in Hawaiian, they wished KP2 well and healing. The ceremony culminated in the offering of a flower lei, a cultural symbol of safe travel, that was draped over the seal's transport cage.

The ceremony was a temporary truce between NOAA and the people of Molokai. News of KP2's blindness had suppressed the divisive storm stirred up by his abrupt removal from Kaunakakai Wharf. Although not happy that the young seal was leaving the islands, everyone finally recognized that neither the veterinary expertise nor surgical facilities to care for KP2's cataracts existed in Hawaii. But Walter Ritte, ever the activist, remained inflexible on one point: KP2 had to be returned to the islands. David Schofield agreed, vowing that KP2 would be brought back within a year—*if* he could find a home for him in the islands.

Less vocal was the smallest attendee, Kahi, KP2's swimming companion from Kaunakakai. He had heard about his friend's diseased eyes and impending move. The young boy decided to make the trip to Oahu to help with the blessing, but mostly to see his friend one more time before it was too late. With only a foot of water in the pool, the walls of KP2's enclosure were too high and the distance too great for the playmates to touch. KP2 splashed expectantly below Kahi. This time, however, the boy did not jump in to play. Instead, Kahi shyly whispered to the seal, "I love you, KP2," and left.

I immediately recognized how Kahi and KP2 were able to read each other and communicate without a word spoken. Others more spiritual than I could see it, too. Years before, a shaman in a Honolulu park outfitted in full headdress had picked me and my dog Austin out of a crowd. Pointing us out to the surrounding people, he'd announced loudly, "You! You have good aura!"

"And him?" I said looking down at my short dog. "What about him?"

"Ah, now, that is what I mean. Together you create a force."

For reasons that could not be explained by science, Kahi and KP2 were drawn to each other in such a force of nature.

Soon the remainder of the participants slowly drifted from the aquarium, leaving the seal alone. One of the last to leave was Walter, who wished the seal luck, ending with, "I wish him well-being and a safe journey—back to Molokai, where he belongs!"

AFTER THE BLESSING EVENT, there was only a handful of days to prepare KP2 for the biggest adventure of his young life. The decision from Washington to allow the transfer of the monk seal from Hawaii to California had never been officially announced. Mysteriously, Amy and Jennifer had prevailed, and the seal's move became fact. There was never any official "yes" to the move; it just seemed that the resistance had grown tired or simply no longer cared about a seal with diminished eyesight that could not be returned to the wild. Relegated to a life in captivity, KP2 was now deemed worthless to the wild population by members of the Hawaiian Monk Seal Recovery Team and the Marine Mammal Commission.

By default KP2 had found a home in my lab.

AS LUCK WOULD HAVE IT, the U.S. Navy was moving one of its Dolphin Systems from a classified mission in New Caledonia back to its base in San Diego. The same Systems dolphins that had once lived at the Kaneohe marine corps facility where I had worked and where KP2 had stacked coral to make an artificial reef now came to his rescue. Dr. Mark Xitco, a colleague of mine from the U.S. Navy's Marine Mammal Program, arranged for KP2 to hitch a ride on the military transport C-17 as it made a fuel stop in Honolulu on its way to the mainland. Miraculously, I was off the hook for a $17,000 FedEx bill.

Nonmilitary personnel were ineligible to receive a security clearance to travel on the military aircraft. KP2, on the other hand, was considered less of a threat in terms of revealing Navy secrets, as long as he could fit through the small side door of the plane. Despite the size of the cargo hold, the aircraft would be crammed with dolphins and their containers. Thus began the first creative task of the seal's journey. How do you move a marine mammal that typically spends its life in the sea (and weighs more than a Great Dane) through the small accessory door of a C-17 for a transoceanic flight? Fortunately, seals are not like fish: they do not need to stay in water in order to breathe. Nor are they like the dolphins on the same flight as KP2, which required large fiberglass tanks partially filled with water and stretchers so that the animals could float while traveling. Instead, a dog carrier would do, if we guaranteed that the seal couldn't break out in the middle of the flight or defecate on the floor and into the electrical wiring. The cargo master was emphatic about that.

On November 24, pilots from the 446th Airlift Wing division of McChord Air Force Base in Tacoma, Washington, landed an enormous C-17 at Hickam Air Force Base on Oahu. The plane, loaded with four mine-detecting dolphins fresh from working in New Caledonia, was stopping only for fuel and KP2. The dolphins at the center of the military transport had been part of an international team of mine hunters on a mission called Lagoon MINEX 2009. During World War II, the Australian military had seeded the waters surrounding New Caledonia with minefields to prevent Japanese ships from reaching island ports where the U.S. had positioned military bases. Almost seventy years later the specially trained Systems dolphins were enlisted to find and destroy more than two hundred contact mines abandoned in the waters around the island. These were professional marine mammals.

The McChord 446th was no newcomer to moving large pieces of military equipment or unusual cargo. Although helicopters, Humvees, and people were the norm, the same group had also provided transport for a wide variety of marine mammals. Their largest and most famous animal passenger was the Humvee-sized Keiko the killer whale (of *Free Willy* fame) on his way to release in Iceland in 1996. Transporting a young monk seal would be comparatively easy, but was given no less military precision. KP2's itinerary was orchestrated to the minute, and involved more personnel than a presidential motorcade to Air Force One.

On the dark edge of dawn, NOAA and NMFS personnel arrived at the Waikiki Aquarium to prep the seal for travel. The motorcade formed by KP2 and his many well-wishers was given a Honolulu police escort to Hickam Air Force Base, and by eight a.m. KP2 was airborne.

Mark Xitco spent the next five hours in the noisy cavernous body of the C-17, shivering and recording the seal's vital signs. For the sake of the marine mammals on board, the cabin temperature was lowered to a chilly 67°F and the cabin pressure maintained at less than five thousand feet. After years of flying Dolphin Systems on military missions across the globe, the U.S. Navy had found that these cabin conditions were critical for transporting animals anatomically designed for ocean living. Because dolphins and seals are built to withstand high hydrostatic pressures when diving, placing them in an airplane at altitude, where pressures were lower, affected their breathing. Just as human divers risk decompression illness on flights immediately after diving, marine mammals were vulnerable to bends-like symptoms during air transport. To avoid any potential respiratory problems, the 446th Airlift Wing from McChord adjusted the cabin pressure to just above "sea level" for their live cargo transport.

A TAILWIND HELPED to slip KP2 across the Pacific Ocean an hour earlier than expected. Awaiting his arrival at the Naval Air Station North Island, in Coronado, California, were Traci, Beau, and me, accompanied by an undergraduate student volunteer and a little yellow Penske rental truck. We had made the 468-mile journey from Santa Cruz to Coronado the night before, and formed a shabby welcome committee compared to the royal Honolulu police escort and send-off on the other end. What we lacked in glamour, we hoped to make up in enthusiasm.

Mark Xitco stepped off the plane with a sheaf of papers in his hands. In his usual brisk, professional manner Mark took me aside and explained the NMFS and Endangered Species Act permits, interstate health certificates, and U.S. Fish and Wildlife Service (USFWS) transfer documents that would allow KP2 to enter California. Walking around the cage, Mark took one last inspection of KP2 while the San Diego–based USFWS agents curiously eyed the paperwork and the tropical animal.

"One last thing," he said. "This seal's official DOD name"—Mark's demeanor suddenly softened with a wide grin—"is Smoodgey."

My lab companions gave each other a series of sideways glances. Apparently, "Smoodgey" had nuzzled his way through Mark's and the transport crew's hardened military shells over the five-hour flight. Unlike the practiced mine-hunting dolphins, the little seal had bounced around his cage, calling to everyone in his throaty Marlon Brando rumble. He had smashed his whiskered muzzle into any offered hand that passed his way. For those on the transport, it had been like traveling with a large, wet Labrador retriever puppy.

Even the most practiced military personnel were not immune to KP2's charms.

8.

LA Landings

With the help of the Navy crew, my lab loaded our rambunctious cargo into the yellow Penske rental truck. Taking a big breath, we headed with dread into the greater metro San Diego and Los Angeles commuter showdown. In the back of the truck, traveling alongside KP2, were Beau and Christina Doll, an aspiring veterinarian in my lab who had volunteered to assist on the transport. They'd set up two plastic Adirondack chairs, a lantern, blankets and tarps, a deck of cards, and a computer with iTunes to while away the hours. Every fifteen minutes they measured KP2's breathing rate using a stopwatch and his skin temperatures using an infrared temperature sensor. The pair was also responsible for dousing the seal with his garden sprayer and jotting notes about the animal's behavior, from sleeping to scratching to rumbling vocalizations. Periodically, they would offer the seal a thawed herring from a red picnic cooler that had been filled with fish and ice from the U.S. Navy dolphins.

Traci drove while I took the comparatively easy role of navigator. Until now, Traci had reserved comment on the arrival of KP2 in my lab. She loved training sea lions and sea otters, was ambivalent about dolphins, and had never thought much of seals. In comparison to other marine mammal species, seals were slow and boring to her. There was a reason that California sea lions were the pinniped of choice for animal shows: they could be trained to leap, hand-stand on their front flippers, salute, balance balls on their nose, and do somersaults in the water. Seals were not that flashy, barely managing to roll over lengthwise like a plump sausage.

While strapping KP2's aluminum cage to the inside truck wall, Traci had taken her first long look at the 120-pound seal that was going to be under her care for the next year.

"Hmph, he's too skinny," she concluded.

"He's sleek," I countered.

KP2 was streamlined and a beautiful silvery gray after molting his puppy coat. His new fur would have been perfect had it not been for two cigarette pack–sized marks on his back. These were the remnants of the satellite and radio tags that had been glued onto his fur by NMFS researchers to track his whereabouts in the waters around Molokai. There was one additional unique feature on KP2's coat that Traci noticed. He had a small white patch of fur on his left thigh. In humans such a light patch of hair would have been called an angel's kiss or a wisdom mark. In keeping with KP2's Hawaiian heritage, Traci called it a tattoo.

. . .

THE TIMING OF KP2'S ARRIVAL could not have been worse: we were driving into the heart of commuter hour on Highway 5 heading into Los Angeles on Thanksgiving Eve. Almost immediately, our yellow truck began drowning in a sea of brake lights on the most congested highway in the nation on the most notoriously congested travel day of the year. Despite the absurdity, everyone in my lab had learned to take such events in stride. Driving with an endangered marine mammal in the middle of Los Angeles was not any odder than putting cameras on Antarctic seals, training mountain lions to run on a treadmill, or carrying the carcass of a stranded dolphin or road-killed coyote in the trunk of a car. Comparative physiology was adventurous and messy. I loved it, and the students and staff working with me accepted the unusual as part of the job.

The animals awaiting us at the marine lab were also unique, although deemed failures in their earlier lives. Puka and Primo from my postdoctoral days at the Navy Dolphins Systems program were afraid to hunt mines. Three resident sea otters, Taylor, Wick, and Morgan, had been expelled from a Monterey Bay Aquarium rehabilitation program when it became clear that they preferred to hop on kayaks rather than sleep in kelp beds. My quirky staff and volunteers who took care of them empathized more with animals than people. Dogs factored big in all our lives. A Saturday night date for the people on my staff as often as not was a drive-in movie snuggled in the back of an SUV with their dogs. My own fit into Santa Cruz was just as implausible. I was a button-down East Coast preppie, educated by nuns, now living among the hemp-smoking West Coast Hare Krishna vegans. A nearly blind, orphaned tropical seal that thought he was human would fit right in.

Fortunately, we had all found each other in the freethinking coastal town of Santa Cruz.

There was one rule in my lab. As heart-wrenching as it could be, I was never so bold as to disobey nature's law on my expeditions or in my research. Over the years, we had to leave orphaned Weddell seal pups to face certain death in Antarctic blizzards. I had stood helplessly by as Adélie penguins bled out from bite wounds inflicted by leopard seals. Neonatal dolphins that were stranded after their pod mates attacked them were euthanized on the beach. Sometimes Mother Nature was an exceptionally harsh taskmaster and I knew better than to question her.

As we navigated through Southern California traffic, I realized that for the first time I had broken my own lab rule. I had relented in the case of KP2 and the Hawaiian monk seals. Instead of abandoning the young seal or his species, I found inspiration in their tenacity. These seals had endured hunting to near extinction during the 1800s, when men sought their oily blubber. Generations of human activities had polluted the monk seal's habitat and fished out their coastal prey. Now the fragmented species was literally rubbing shoulders with humans in the main Hawaiian Islands in order to survive. We owed it to their species to bring their numbers back. KP2 was the first step.

"Hey, we've got a blind endangered species here!" I shouted at the silver Mercedes that cut off our rental truck. The sudden braking had initiated a series of loud crashes on the fiberglass wall behind me. Traci quickly got on a walkie-talkie.

"You guys okay back there?"

"Mmmph," was the muffled, static-filled reply.

"They're fine," Traci concluded confidently. My trainer was young and unflappable, and could read animal behavior better than anyone I knew. Without a doubt, Traci was one of the best marine mammal trainers in the business. She took no guff from animal or man alike. Few crossed her after seeing how she was able to humble the three snappy, snarly sea otters into kittens. On this adventure I was quickly learning that she was fearless on the road, as she maneuvered our boxy truck into the commuter lane to the honking displeasure of the frazzled LA-bound drivers.

"What if the commuters surrounding us knew that there was one of the last Hawaiian monk seals on the planet edging past them in the diamond lane?" I asked Traci.

"I doubt they would drive any differently."

As commuters aggressively swerved, braked, cursed, and honked, KP2 remained nonplussed, yawning and sleeping to the music of Taylor Swift, Garth Brooks, and Carrie Underwood in the back of the truck. Beau and Christina had discovered that the twang of country-western tunes caused the seal's large brown eyes to begin drifting, and soon set him to snoring.

Although the truck bed was rapidly cooling as we headed north, Beau, like any self-respecting NorCal surfer, would not succumb to wearing long pants. He had weathered Antarctic blizzards in his board shorts and T-shirts—even if they were under a down-insulated parka and coveralls. He was not about to abandon the uniform of Santa Cruz while barreling down the highways of his home state. Despite her suntanned blond surfer girl looks, Christina had no such reservations and wrapped herself in a wool blanket to ward off the chill night air during the long drive north. They draped a blanket over KP2's cage to keep his Hawaii-acclimated body warm.

WE INCHED OUR WAY with thousands of other pre-Thanksgiving travelers along the highway. Signs for San Diego, La Jolla, Del Mar,

Encinitas, Oceanside, and Camp Pendleton drifted by. Eventually, Traci pulled into a rest stop less than eight miles from the highway border patrol checkpoint so we could check on our passengers. Undoubtedly, this was the last bailout stop for illegal immigrants who had been smuggled across the U.S.-Mexican border.

"How's it going?" I asked Beau and Christina as Traci rolled up the back door of the truck. The stench of wet seal and fish was overpowering.

"No problems here," Beau responded cheerily. Despite the lack of sleep, the trainer was as bouncy as the seal.

"Pretty smelly ride, eh?" I observed, sheepishly aware that at least Traci and I had fresh air in the front cab.

"What smell?" Beau honestly seemed incredulous. Thank goodness for the hardened noses of animal trainers, I thought as we relocked the truck door and slipped back onto the highway.

Traffic was now at a complete standstill. We rolled forward at a snail's pace past the state weigh station, and the San Onofre and San Clemente exits. It was not until we had passed the exit for San Juan Capistrano, fifteen miles past the normal border patrol stop, that Traci and I suddenly realized that there had been no checkpoint. The border patrol had taken mercy on the holiday travelers; there would be no inspections this evening. We were spared the awkwardness of explaining our immigrant cargo to suspicious border agents. Our seal was safe.

DINNERTIME CAME AND WENT as Traci bulldogged our truck through the tangle of freeways bisecting Los Angeles. While KP2 had been able to dine alfresco, our stomachs began to rumble. In desperation, Traci maneuvered the yellow truck into the bastion of fine California highway cuisine, In-N-Out Burger.

The parking lot was packed with highway-traveling dog owners who'd had the same idea. Unbeknownst to us, the lawn of the fast-food establishment was a favorite pit stop for people traveling with pets. While the owners slid down burgers and fries, dogs of all sizes and shapes squatted and leg-lifted on the landscaping.

I took up seal guard duty in the back of the truck as the others went to order food. It was the first time that I was alone with KP2. We gazed at each other with uncertainty in the awkward silence of a first date. The seal watched my hands and eyes. I studied him, too. If I moved the garden sprayer, he automatically rolled over for a dousing. If I looked at the red cooler, he followed my gaze and blinked expectantly for a fish. He responded to my subtle moves, almost predicting my next act before I even knew I would make it. I instinctively responded to his behavior in the same way as if caught in a dance.

From a distance, I heard a low thumping and immediately thought, "Earthquake!" However, KP2 would not fool me a second time. I ignored the steady rumbling and the thumping buildup, and prepared for the final outburst of *"Brrawwrrrr!"* Turning to the passing dog owners, I smiled and secretly wished that KP2 would quiet down.

"Rraaaughhrr." The seal pup called loudly in his Brando growl. In all of the hours on the road, I had never heard the seal make a sound. Now he wouldn't stop. KP2's renewed activity soon caught the attention of the dog owners in the parking lot.

"Rraaarugh!" KP2 repeated. The throaty thumps were unlike any marine mammal sound, unlike any seal vocalization I had ever heard. He crawled around his cage and was not interested in the fish I offered from his cooler. In the yellow and red neon glow of the In-N-Out

Burger sign, I had the distinct feeling that the seal was testing me. He waited expectedly for a hand to smash his muzzle into. Instead I kept my distance, refusing to give in to his cuteness.

A moment later the others returned. As the four of us ate our meal on the Penske tailgate, in the dark of the autumn evening, people and dogs streamed by, gawking at the interior of our vehicle, not sure what to make of our peculiar pet. The seal's rumbling *"rrrraaaaughhs"* were most un-canine. We ignored their stares and ate our dinners nonchalantly, as if our legless, earless companion were some type of weird pit bull.

SATIATED AND OUT OF THE CITY, we hoped for smooth sailing for the remaining 350 miles to Santa Cruz. The last challenge was a rugged, twisting segment of Interstate 5 known as the Grapevine connection. This section of highway rises out of Los Angeles into the surrounding Tehachapi Mountains. Notorious for snow, fog, and car pileups, the Grapevine is 4,183 feet at its peak and has a dramatic, tortuous 6 percent grade on the descent into the vineyards and agricultural fields of California's Central Valley.

The veterinarian at Long Marine Lab, Dr. Dave Casper, had insisted that we monitor KP2's breathing as we made our way through the pass. More curious than concerned, he wanted to know the lower limits of altitude that marine mammals could travel without a change in breathing patterns. Dr. Casper had read about respiratory complications experienced by sea otters and sea lions transported in small aircraft over these same mountains. Was it the speed of altitude change or the actual altitude level that instigated faulty breathing? Besides curiosity, the veterinarian was also not taking any chances with our

Hawaiian cargo. He did not want KP2's first breaths of mainland air to be his last.

"Okay, guys," I instructed Beau and Christina. "Every five minutes, write down the breathing rate of KP2. Call us on the radio if there is a problem." The seal's companions dutifully took up their positions in the Adirondack chairs and settled in for the final stretch of the ride north.

Inexplicably, Cal Highway had chosen the Thanksgiving holiday rush to repair two of the three lanes of the Grapevine connection. Bright orange road construction signs and traffic cones forced the traffic to slow down on the treacherous stretch of road. We were squeezed with holiday travelers into one long brake light that snaked over the pass. Our truck crawled uphill in a low-gear hum. After nearly an hour, our headlights peaked over the top of the ridge with no complaint from the vehicle nor the passengers traveling in the back. Just as we started to tip downward, the highway lanes opened up, causing vehicles to begin careening down the mountainside.

Traci had just begun to pick up speed when there was a sudden bang in the back of the truck. Chairs crashed and bodies shifted quickly, causing the truck to sway.

"Ow, wow!"

I tried to grab the walkie-talkie while Traci steered.

"What's going on back there?" she yelled.

"*Earchhhh!*" Beau replied. We could hear Christina groaning in the background.

"What? Repeat!" Traci cursed the radio.

Silence.

Caught on the downhill curves, our truck quickly accelerated. We were trapped in the fast lane as big rigs blocked the middle and slow lanes in a last-ditch attempt to make their holiday deliveries. There was no choice. We had to let gravity take over. Whatever emergency was happening in the back of the truck would have to wait until we pulled out of the Grapevine. I sweated the downhill miles, convinced that KP2 was in respiratory failure and that I had made a horrible mistake taking the little seal out of Hawaii.

Through the back wall of the truck there was another round of yelping, then moaning. Traci and I navigated into the Central Valley, trying not to let our imaginations run away with the speed of our descent. We had just reached the bottom of the hill when Beau's voice finally crackled over the walkie-talkie, "Nuclear attack!"

Before I could respond, Traci burst out laughing.

"Gas explosion," she explained.

"You're not kidding!" Beau shouted. "We were bombed. Repeatedly!"

"Must have been the altitude change," Traci yelled over the static of the walkie-talkie. "Suck it up! It will clear out in a few miles."

"Consider yourselves lucky," I added. "You've already learned something new about Hawaiian monk seals." Like little old men on airplanes, changes in altitude and pressure made monk seals fart.

It was o-three-hundred, in military terms, by the time our little yellow truck finally rolled into Long Marine Lab at the University of California, Santa Cruz. With considerably less fanfare than his early morning send-off, the team carried KP2 and his cage under the cover of darkness into the holding pool area. At long last, KP2's cage door was opened. Then we waited, and waited some more. KP2 continued to sleep with his head smashed in the far corner of the cage, oblivious to his freedom.

Like a child hauled over too many time zones, the little seal was in no mood to move. Twenty-one hours of travel had taken the spring out of him. Encouragement from the handful of student volunteers who had waited all night for his arrival finally sparked his curiosity. The seal inchwormed his way out of the cage; in sluglike movements he circumnavigated his new pool. Following a low snort, he dunked his head into the water, and then fell in with an awkward splash.

WITHOUT EVER TAKING a single swimming stroke, KP2 had traveled more miles across the Pacific Ocean in one day than his wild family would ever undertake in a lifetime. He deserved the lazy entrance, if less than graceful, into his outdoor pool.

As we looked on, one of the volunteers asked, "What's his name?"

I hesitated. The seal was entering a new phase of his life, and it was time to begin using his grown-up name. During the blessing ceremony the week before, KP2 had been given an official Hawaiian name chosen by the elders of Molokai. He was christened Ho'ailona. "It means a special seal with a special purpose," Walter Ritte had explained at the time.

"Ho'ailona," I replied to the student. "His name is Ho'ailona: a sign from the ocean." And so with that early-morning introduction, a special Hawaiian monk seal with a special purpose arrived at college in Santa Cruz. Almost immediately he proceeded to turn my world upside down.

9.

Sealebrity

While many biologists study the smallest cells, molecules, flies, and mice, I chose instead to focus on the biology of the biggest animals on earth. Seals, dolphins, whales, lions, sea otters, elephants, and other large mammals that weigh between thirty and thirty thousand pounds are my research subjects. This range of body sizes is specific, for it comprises the group of animals most vulnerable to human perturbation. Long life spans that could exceed two hundred years in the case of some whales, low reproductive rates with many species giving birth to only one offspring per year, and enormous environmental resource needs make big mammals exceptionally susceptible to the effects of pollution, climate change, and habitat destruction.

Big mammal science is timely, exciting, and challenging. It is also the most difficult to conduct, particularly with endangered species, which are nearly impossible to study. KP2, the nearly blind problem child of NMFS, the surplus male of a dying species, was now the most important addition to my physiology program.

My lab is a zoological school, a place of wild aspirations. It is an institution where young students can live their dream—if only in college—of saving the world's animals. Santa Cruz, a Northern California town still clinging to seventies-era idealism, is the perfect environment. Locals, including those running the university campus, are radical enough to believe that such ambitions are not only possible but necessary.

Together, the members of my lab strive to discover the unique vulnerabilities and capabilities that make a dolphin a dolphin, an otter an otter, a seal a seal. The underlying logic is simple: humans have to know *what* they are saving before they are able to save it. Thus, the animals and humans of my lab work side by side, day after day, in the singular goal of trying to understand and preserve wild species. Science paves our road to conservation.

With KP2 in residence, however, it would not take long for us to discover just how difficult saving an endangered species could be.

EVEN WITHOUT AN official announcement or press release, word leaked out that a celebrity seal pup was hidden behind the wooden fencing at the University of California's Long Marine Lab. A wave of curiosity seekers descended, all demanding a piece of the exotic visitor from Hawaii who had made the front page of the *Wall Street Journal* and enjoyed a YouTube following. Overnight, reporters from television stations and newspapers from across the states began choking my phone line with requests, a situation that quickly devolved from amusing to appalling.

Pinniped paparazzi, journalists jockeying for an exclusive interview and pictures of KP2, arrived with the grace of a rogue tsunami. They caught my lab unaware and wholly unprepared. I was especially naive, having lived the blessedly isolated life of a wildlife biologist. Except for Nobel laureates and icons such as Jane Goodall, scientists are not usual fodder for paparazzi.

Ignoring reporters' calls only escalated their intensity. Refusals ended in a demand to know "What are you hiding?" and the threat of "crash" visits. This was not how I envisioned saving animals.

The media frenzy continued into the dark of the night several days after KP2's arrival when an intruder attempted to climb over the marine lab's security fence and force his way into the marine mammal compound. One of the student volunteers discovered splintered wood and a broken door handle on the ground the next morning where the person had hoisted himself up.

I grumbled while toeing the debris on the gravel. "What did they want? Pictures? To steal KP2? To hurt him?" I did not understand what motivated some people to do the crazy things they did to animals. Scientists were so often cast in a negative light when it came to the furred and finned. In my experience, biologists were often the ones left picking up the pieces when someone harmed an animal.

One such incident had happened when I was working in Kaneohe. During the middle of the night a group broke into the U.S. Navy's Dolphin Systems compound. In the morning the trainers and I discovered one of the bottlenose dolphins acting skittish and floating in its pen. On closer examination we found several open wounds along its body. Thinking that the animal had inadvertently swum into a sharp object, one marine mammal trainer dove into the pen but could not even find a stick that would have poked the dolphin or created such clean wounds.

Further investigation around the compound ultimately uncovered a drunken episode. Apparently after a party, several men had swum around the point that separated the military base from the dolphin holding area. Stumbling onto the wooden pier, one of the inebriated men had tried to pet the dolphin. When his advances were refused by the suspicious animal, he had taken a knife and stabbed it. Fortunately, the wounds were superficial—a dull blade is no match for tough dolphin blubber—and we were able to quickly heal the dolphin physically if not psychologically. The young man did not fare as well. Before completely recovering from his hangover, he was arrested for destruction of government property, which in this case was a federal offense involving prison time.

STANDING ON A LADDER and peering over the edge of the fence, I could see that there was no damage to KP2's enclosure. The chain-link and wooden fencing around his cement deck and run was intact and the water of the small inground pool was clear. "I don't think they made it past the top," I shouted down to Traci. "It looks like whoever it was fell backwards." Taking in the ten-foot drop to the ground, I added, "I hope it hurt!"

"Nothing was moved inside of KP2's pool. No one made it in," Traci concurred, shaking her head. "I only wish I had seen them."

I wished she had seen them, too. I relished the thought of any intruder who had the misfortune of encountering Traci during her watch. She had the maternal instincts of a wild tiger when it came to protecting the animals in her care. Because she was female, there was the mistaken notion that she was a pushover. In reality, Traci had an athletic, and at times pugnacious, kickboxer edge when challenged. While Beau was the laid-back island boy, my female trainer packed a mean punch.

"Let's clean the place up," she directed her circle of volunteer helpers.

No sooner had I finished dealing with the damage from the night-time break-in than the intrusion became personal. The light on my office phone blinked angrily for attention. A series of messages from my campus bosses and the NMFS Office of Protected Resources all demanded the same thing: respond to a persistent journalism student who wanted private access to KP2.

The previous day the same student and I had clashed on the telephone. She had not taken my refusal well, the identical refusal tendered to all callers. "Who are *you*?" she demanded. She immediately called the UCSC administration and my NMFS permit officers.

"Your permit is being FOIA'd," Amy Sloan from NMFS informed me. Using the Freedom of Information Act, the journalism student had contacted the government agency for access to my records and the young seal. She sought out the public relations office for my university, wanting information about who was in charge of KP2. Now my job and marine mammal research permits—the lifeline of my work and fifteen years of applications—were in jeopardy. Fear of lawsuits by the school and the government upped the pressure on me to relent. The resulting problems were enormous, and I resented the time taken away from my science. Still I refused.

"I'm sorry. This should not have happened," Amy apologized over the phone. She began to regret having helped to instigate the move of KP2 to my lab. "Maybe Jennifer and I should not have gotten you involved."

"Never think that!" I countered emphatically. "I knew the risk. Besides, if things get too legal I'll go to Africa and work on lions..."

While Amy dealt with the FOIA request, I maintained my hardline stance at the lab. Backed against the wall, I did as any wild fox might. I flatly refused to budge. I snarled a curt "No!" to all journalists. FOIA be damned.

My unwavering refusal to grant interviews to the journalism student and the rest of the media was driven by my worst fear for KP2: disease. With each request from a stranger to enter into KP2's enclosure, images from Molokai's history flashed before my eyes. To me, all strangers carried an invisible threat to the young seal.

A family of microscopic organisms with disproportionately large names, the bacteria *Mycobacterium leprae* and *Mycobacterium lepromatosis*, was the source of Hansen's disease that had devastated Hawaii. The bacteria had been unwittingly carried by strangers to the islands, Chinese laborers who had arrived to work in the sugarcane fields during the early to mid-1800s. Even a proud warrior culture like the native Hawaiians' had no defense against the microbial invasion. Contact with the bacteria was disastrous, resulting in the epidemic that swept through the islands by 1863. With no cure available, the only control was complete isolation of those fallen ill with the disease; hence the development of the quarantine colony on Kalaupapa Peninsula.

In part, this vulnerability to invasive diseases was the hidden price for living in tropical paradise in the middle of the Pacific Ocean. Residents of Hawaii reside approximately 1,860 miles from any other continent, and have the distinction of being the most geographically isolated community on earth. With this distinction comes a naive immune system unfamiliar, and therefore exceptionally vulnerable, to diseases carried by outsiders.

Airplanes and the constant influx of tourists have altered the situation for modern Hawaiians. Not so for KP2 and the monk seals. Like their historical human counterparts, seals isolated in the outer Northwestern Hawaiian Islands and in the surrounding waters remain immuno-naive.

As the first Hawaiian monk seal pup to breathe mainland air, KP2 was now encountering an invisible soup of viruses and bacteria unlike any in Hawaii. They, too, came with mysterious scientific names: *Lyssavirus* (rabies), *Flavivirus* (West Nile virus), and a member of the *Paramyxovirinae* (phocine distemper virus). There was no predicting KP2's susceptibility to these diseases, which seemed easily tolerated by the local harbor seals and elephant seals. There was no predicting the ability of humans to inadvertently carry the microorganisms on their hands, their clothes, or their breath. As a result, KP2 had to be treated like any newborn, and safeguarded until his immune system had time to slowly respond to the jungle of mainland microorganisms.

KP2's health also had to be guaranteed without flaw before he could be exposed to the other animals of Long Marine Lab, or they exposed to him. His lab neighbors included bottlenose dolphins, sea lions, sea otters, and seals involved in a wide variety of research projects. Scientists at the lab were trying to decipher the effects of oceanic noise on marine mammals as well as employ the animals as models for endangered killer whales, Steller sea lions, and marine otters. Soon all of these valuable, specially trained lab occupants would be sharing the same seawater and salty air with KP2; their health was paramount.

Like the turn-of-the-century Hawaiians of Molokai, KP2 was placed in strict quarantine to safeguard him and the animals around him. By federal mandate, he had to remain in isolation for sixty days. If he showed no signs of illness during that period and passed a final veterinary examination, then he could be released into the main pools. Until that time only a handful of caretakers were allowed to contact him. I was the gatekeeper, and I banned all journalists.

QUARANTINE LIFE FOR KP2 was not all that different from his days at the Kewalo Research Facility or the Waikiki Aquarium. We provided him with a personal warm pool with a skylight roof and sunning deck complete with daily cleaning service supervised by Traci and Beau. Meals were delivered on demand with a wide selection of fresh seafood. For his first two months in California the endangered seal resided in a private sanctuary with all the amenities of the Four Seasons resort up the coast.

"The only thing missing is a mint on the pillow," I noted to Beau.

"Oh, don't worry. We offer him a good-night herring each evening." Beau was looking ragged. The excitement of having an exotic seal had been distilled down to the reality of taking care of him. Traci, too, had spent weeks creating the heated enclosure before KP2's arrival and then driven across the state with the seal. My team was exhausted before we even got started with the science.

An animal in quarantine created a whole new level of complexity for animal care in our small facility. The law required us to maintain the sterility of a hospital in the middle of the chaos of an active research station. At times individual missions chafed.

With a single food preparation kitchen for the dolphins, seals, sea otters, and sea lions as well as common walkways and limited staff, quarantine meant creative scheduling. Personnel working with KP2 or his food had to change clothes and take showers before they could move between quarantine and nonquarantine animal areas. Shoes and boots had to be dipped in plastic dishpans filled with disinfectant solution when entering or leaving KP2's enclosure. Hands were washed incessantly to the point of rawness. Ultracleanliness was critical.

Around the lab it quickly became apparent who was working with me and KP2. Our hands were red and chapped from too many washings. Clothes were discolored and tattered from the ravages of bleach used to clean his pool and deck. Our shoes were stained and warped from too many disinfectant dippings.

Despite the added workload, my team arrived each morning shivering in the anticipation of being a part of something new and exciting. Traci, Beau, the undergraduate student volunteers, and I viewed KP2 as the last of a dying tribe, a symbol of the human footprint on the seas. There was a childlike wonder when the caretakers met KP2, and a sense of awe at being able to reach out and touch this member of an endangered species. With the same innocence that had inspired me to seek salvation for the rescued forest animals I brought into church as a young girl, my team hoped to save this one precious ocean creature.

For that reason, I rabidly defended KP2's privacy and my team.

When I reported the lab break-in attempt to the university veterinarian, I asked him, "What is the matter with people?" Dr. Dave Casper was now the lead veterinary caretaker for KP2, a job that included serving as amateur psychoanalyst for the humans working around the seal. He was the same university vet who had warned us about the dangers of altitude changes when transporting seals. I wished that he had warned me about the human element.

Balding and bespectacled, Dr. Casper had earned his veterinary stripes through a long history of working with the largest marine whales in national aquariums as well as treating the smallest of pets in private practice. He moved easily between the care of the wild and exotic, and the intimate relationships of pet owners with their pets. Few realized that Dr. Casper's country-doctor shell concealed a highly inquisitive computer geek who loved modern digital veterinary medicine. He was always testing me on some new molecular theory or physiological-environmental interconnection for wild animals. The veterinarian was also an enthusiastic observer of the human animal.

"Well, what did you expect?" Dr. Casper laughed at my frustration with the journalists. "You've got Elvis of the Seals in your hands. Better not screw up!"

BUT I WAS SCREWING UP. Less than a week in and I had a much bigger problem than journalists on my hands. KP2 refused to eat.

Trying to get calories into him had turned into a war of wills between the seal and his trainers. After months of eating whenever and whatever he wanted in the waters surrounding Molokai, KP2 now turned his nose up at the proverbial California broccoli we offered. We had plump herring, juicy squid, and lovely smelt that all the other animals ate with relish. KP2 sniffed at them all in disgust.

The young seal displayed his dietary opinions openly on his muzzle and in his cloudy eyes. With every fish offered by Traci and Beau, KP2 screwed up his whiskered upper lip and squinted. Reluctantly, with eyes shut and lips pursed tightly, he would slightly open his mouth if the fish was pressing to get in. Ultimately, he'd obstinately turn his head away, leaving the dead fish dangling in the trainer's hand.

Rather than eat, KP2 spent his initial days on the mainland hauled out on his deck in the weak California winter sun with a rumbling empty stomach. He seemed no better off than when he lay abandoned and starving on Kauai.

Concealing the doubt churning in the pit of my own stomach, I decided we needed to weigh our stubborn seal. Gone were the days when KP2 could be weighed in a Rubbermaid container on a bathroom scale. The young monk seal was now over four feet long and would be left with his head and flippers hanging over any conventional scale. We had to carry in a large metal veterinary platform scale that he could lie on.

The moment the scale appeared, KP2's attitude changed. Immediately, the hungry seal perked up. Driven by curiosity and a chance to play, KP2 hopped out of his pool and his funk. He inchwormed his way over to the scale, leaving a skid trail of water on the cement of his enclosure. With little more than a few hand signals from Beau to guide him, KP2 eagerly flopped his wet body onto the metal platform and proceeded to lie still while we recorded his weight.

His enthusiasm was duly noted in glowing terms in his medical records. "Wow, a voluntary weight on his second day here . . . pretty cool!" was the behavioral note written in his daily chart. I had to admit that I was impressed. The simple act of weighing a wild animal is a remarkable feat. Imagine trying to weigh a wild bobcat or lion. People would need to dance around the animal to position it just right on a scale. Then the animal would have to sit perfectly still while the scale stabilized. Many zoos and aquariums don't weigh their animals on a daily basis for this reason. Large wild mammals don't like to lie still, especially in the presence of humans.

Beau had met this training challenge head-on with many species of animals and knew their tricks. Nowhere was this more impressive than at the University of Hawaii vivarium, where he had watched animal technicians trying to net scurrying owl monkeys for a weigh-in. Monkeys were flying across their enclosures as the technicians tried to snare the evasive animals. When he suggested training the monkeys to weigh themselves, the technicians had laughed. Proving them wrong became a quest. Within several weeks Beau and his team not only had the monkeys standing quietly on the scales for weighing, he had them returning to their cages and locking the doors behind them. The technicians were ecstatic until they found that in the process of learning how the door locks operated, the monkeys now also knew how to reopen their cages and could escape with ease. Worse, there was a real fear that the monkeys could let their friends out. Sometimes a little knowledge was a dangerous thing.

KP2 figured out the weighing task in one session with nary a word from any of us. He obediently stayed in his pool while the digital scale was set up. Then, when Beau was ready, the seal flopped right onto the scale without hesitation. He had even posed as if a snapshot were being taken while the scale stabilized.

"That's one smart seal!" Ashley, one of the student volunteers, remarked.

"Yes, but if he is so smart, why won't he eat?" I asked in return. There had to be a reason behind his refusal. The only negative part of KP2's first weigh-in was his fish drop. Beau had praised the seal for his excellent behavior as he sat on the scale. But when the trainer handed KP2 the equivalent of a "good dog" biscuit—a fatty herring tail that is the candy of fish for marine mammals—the seal let it fall to the ground uneaten, and slunk back to his pool.

I caught my breath when I recorded his weight. The digital numbers on the scale settled into a final weight of 54.4 kilograms (120 pounds). Although KP2 was only seventeen months old and nearly weighed more than me, he was underweight for a monk seal of his age. Traci was right; he was skinny. His body was beginning to reflect the ravages of his unorthodox early days and his current refusal to eat. A bad mother and a scavenging lifestyle in the shallow waters around Molokai had whittled his body down. Running my hand down his slick back, I could feel the bumps of vertebra along his spine. Hip bones and shoulder bones were starting to protrude, giving him an angular appearance that made his fur drape. Instead of the fine, stream-lined, neckless profile of wild seals, KP2 was verging on the "peanut head" look of a starving marine mammal.

I sighed and watched the sun set in a westward blaze below a horizon that connected me to the monk seal's island home. Giving up for the day, I walked past KP2's pools and climbed down the cliff to the beach. I left my running shoes on the cold sand to dip a toe into the sunset-splashed water. I shivered with the initial shock of cold.

Only water separated the people of Hawaii and me, but more than ever I felt that we were oceans apart.

10.

Mele Kalikimaka

Christmas was coming and Pele, the Hawaiian goddess of fire and mischief, proved that she was not yet ready to give up her hold on KP2. To remind him that he was indeed in a foreign land, she delivered a severe Arctic winter punch—the coldest in Santa Cruz history—from Santa's backyard. The seal that had frolicked with bikini-clad surfers in Molokai only two months before was now pelted with ice when he ventured into the uncovered outdoor run of his quarantine area. The surrounding Santa Cruz Mountains turned white with snow, and records broke as the weather service reported temperatures dropping ten degrees below normal. Neither KP2 nor my team knew what to make of the unusual cold.

The tropical seal didn't appreciate the sting of the cold on his flippers and face. He shook his head and shivered. He blinked at the freezing rain and finally, in a move reminiscent of his Antarctic seal cousins, attempted to avoid the repulsive chill by slinking from the run area and slipping underwater in his warm-water pool.

Fortunately, my team had anticipated the onset of winter, if not record-breaking cold, for our Hawaiian visitor. Before KP2's arrival, Traci, Beau, and I had placed a tall order with Randolph Skrovan, the Long Marine Lab facilities engineer.

"You say you want eighty-degree salt water *and* eighty-degree humidified air for a *seal*?" Randolph had responded incredulously. He'd pondered our unusual request for a heated marine mammal enclosure by running a hand through his brushy hair. All of the other pinnipeds at the lab required cold water; the need for a seal spa was a new concept for him. During the winter, heated ocean water was in short supply at the lab as coastal water temperatures dropped into the low fifties. Anyone foolish enough to go swimming or surfing in Santa Cruz from December to March risked wooden, numbed feet and an ice cream headache. Our request to create a Hawaiian environment in the middle of Santa Cruz winter quickly caused Randolph to develop a brown mat of morning hair. Yet I had faith in him.

As a graduate student in my lab, Randolph had once repaired a broken outboard motor with a hood cord cut from his jacket in time to keep a dinghy full of researchers from crashing into a larger ship in the middle of the Aleutian Islands. Then, when stranded shoeless on a remote island after his disabled research vessel had crashed into a wall of rocks, he'd fashioned flip-flops out of a plastic float that he found washed up on the beach. While wearing his self-crafted maritime pink flops, he'd scoured the island for berries and water to sustain himself and the group until the U.S. Coast Guard arrived. Randolph was the ultimate survivor now employed as the ultimate engineer for a marine lab riddled with budget cuts. It was only a matter of time before he engineered a creative solution for the tropical seal.

"You know, there are these dilapidated greenhouses I saw down the road." Randolph's speech revved up. "If we scavenge the plastic and some of the metal support beams, we could create a sorta solarium around KP2's pool. A roof, some siding. Yes, I think that would do it."

Traci helped Randolph design the seal's enclosure and dismantle the abandoned greenhouse. With the scavenged building supplies in hand, they assembled a metal skeleton over KP2's pool and deck area. Cutting large sheets of plastic from the old greenhouse roof, they formed a circus dome and insulating walls. The siding rolled up to allow for temperature control during the day, and then could be secured to stripping during the night to create a warm microenvironment for the seal.

Air in the home-crafted, solar energy–powered "sealarium" heated up. By coaxing the lab's cranky saltwater heaters, Traci managed to bring KP2's pool up to a sultry 85°F to rival the lapping waters of the Kauai beach where the seal had been born. Because the suddenness and ferocity of the cold snap overwhelmed the sealarium, Traci purchased an infrared heat lamp to provide additional warmth during the freezing nights and cloudy days.

The heated pool and lamp were welcome sources of pleasure for the little monk seal during the unusual Arctic blast. Harkening back to his island sunbathing habits, KP2 spent his time lounging on the deck of his sealarium beneath the heat lamp as if in a tanning booth. First he'd toast his hind end. Then he'd roll over to warm his belly, only to roll bottom up again while his trainers and I stood in the outside run area in the freezing rain.

I had to smile, although I knew there would be hell to pay in the end. Eventually, the university accountants would piece together the cause of the sudden skyrocketing electric bill for the lab pool heaters. Until then, I let KP2 enjoy his bit of Hawaii while I lived on the edge, reminding myself, *Only those who attempt the absurd...*

SNUG IN HIS HEATED ENCLOSURE, KP2 was oblivious to the weather and my presence when I silently slipped inside one day, grateful for the warmth. For the first time I had a chance to admire him. He was different from all the other species of seals that I had studied, sleeker in body with a silvery sheen to his pelt. Water shimmered when he swam, leaving me momentarily spellbound. History suggests that corpulent manatees, dugongs, and sea cows were the origin of the mermaid myth. I think not. Watching KP2, I saw grace in his glide and splendor in the way the sun played off his glistening wet back. Surely the sensuous beauty of monk seals would not have been lost on ancient mariners. Had I the power to design a mermaid, I would have started with this beautiful seal.

I watched as KP2 eased himself onto the deck and surveyed the ground with his cloudy eyes. His enclosure was littered with pieces of old fire hoses, balls, and deflated plastic floats that the trainers provided as toys. Fully expecting him to clear a path through the debris, I was surprised to see KP2 do just the opposite. He headed for the nearest pile of toys and flopped on top of it. The seal had all the room in the world and yet chose to cuddle with junk lying on the ground. He hugged a deflated plastic float to his chest. He rolled on top of the fire hose trying to bury his head. Had it been a fishing net, he would have become hopelessly entangled.

"What an odd seal," I remarked to myself. KP2 had still failed to notice me standing in his sealarium. He continued to wrestle with the hosing until he'd created a bed. Then he promptly shut his eyes and fell asleep with his head lolling upside down.

I found the similarity in KP2's behavior to that of wild Hawaiian monk seals remarkable. Unlike skittish harbor seals and placid Weddell seals, which tended to avoid beach trash, Hawaiian monk seals were inexplicably attracted to the flotsam and jetsam that washed up on the island shores. Any piece of rope, fishing nets and lines, plastic bags and floats were candidates for playthings and bedding. Walk any beach with miles of white pristine sand on French Frigate Shoals or Laysan Island in the Northwestern Hawaiian Islands, and you'll find a monk seal draped around the only piece of high-tide garbage in sight. No one knows why. Sadly, oceanic trash from across the globe washes up almost daily on these remote islands.

"But how did *you* figure that out?" I asked the snoring seal. Without the benefit of a mother, father, or siblings to teach him, KP2 instinctively climbed on top of the old hoses and deflated floats. With his chest facing the sun and his head hanging awkwardly down, he fell asleep in the same quirky repose of a wild monk seal.

I named KP2's sleeping position the "homeless pose" as he snoozed atop his belongings. With the sudden cold spell, the sidewalks of Santa Cruz were inundated with an eclectic subculture from the 1970s peace generation trying to keep warm. Like KP2, the abandoned, the downtrodden, the Vietnam vets, and the perpetual hippies had landed in the coastal town anticipating California sun. Instead they received the chill of their lives. They learned to stave off the cold by bundling cardboard or other collected junk between their bodies and the frozen doorjambs.

After watching KP2, I began to wonder if Hawaiian monk seals were attracted to garbage for the same thermal reasons. Despite a substantial blubber layer, a pinniped transfers significant amounts of heat through its skin just like a human lying on cold cement. In one study, California sea lions lost over 25 percent of their body heat from the contact of their bellies on the sand. Even in the Antarctic, where retaining heat is at a premium, Weddell seals sleeping on the sea ice lose so much body heat that they create snow angel outlines where they've been lying. Over several hours of slumbering in one position, these polar seals eventually melt into a bathtub of slush below their hot bodies. It seemed peculiar that a Hawaiian seal would need to find ways to retain body heat, but I made a mental note to test my homeless-seal theory once KP2 was out of quarantine. For now I just watched him doze peaceably on the bed of his own making.

EACH MORNING THE PLASTIC SIDING on KP2's sealarium fogged up as steam rose from his heated pool and clashed with the outside December chill. It was as irresistible as a frosted car windshield, so lab staff and volunteers took the opportunity to draw tropical Christmas scenes for the seal. Outlines of palm trees, Christmas trees, snowflakes, and "Alohas" greeted the seal as he arose.

The decorations did little to prevent KP2's first mainland holiday season from worsening. He had been pelted with ice and offered little in the way of fish that stimulated his appetite. We had also unknowingly exacerbated his discomfort in our rush to create a quarantined, insulated micro-Hawaii for him. Despite our best efforts, I had overlooked the one thing that he really needed in his enclosure.

I discovered this while quietly watching his ramblings one day. With his poor eyesight, KP2 devised alternative ways of sensing his surroundings. His head cocked, he listened intently to the sounds of the marine lab using two pinhole ears on either side of his head. Like other phocids, he possessed no external ears, and instead tilted his head from side to side to localize the source of voices or noise. Motion was also important. Although he had difficulty making out details, he could still detect the presence of someone from the shifting shadows of their movements.

As such, the play of light and the symphony of lab sounds were important sensory cues that kept the visually impaired seal in touch with his environment. We had mistakenly muffled KP2 when we built his heated sealarium. Insulated but isolated, our seal was trapped like the plastic figurines in a Christmas snow globe in which the fake scenery never changes.

Alone in his enclosure, the monk seal constantly listened for any tiny sound through a knothole in the lower section of the wooden fencing. If he detected something interesting, he would then shift his head to peer through the hole with one eye, straining to be a part of the action on the other side.

The swing of his enclosure door or a visitor walking onto the deck caused him to surf his way across the water of his small pool in a frantic greeting. Riding his own speed-generated wave, KP2 would slide onto the deck and then up our pant legs. He identified each of us in a ritualistic series of snuffles up and down our legs, nearly tripping us in the process. Soon he learned to block doors with his body just to keep us with him a little longer. The same people-friendly behavior that had initiated his removal from Kaunakakai Wharf in Molokai made the little seal desperate for human companionship.

Isolation, not weather, had sent him into depression.

"I'LL TAKE CARE OF KP2 for the holidays," Traci volunteered one day. To give Beau a chance to return to Hawaii to celebrate with friends, she offered to take the brunt of the feeding schedule for all of the animals at the lab.

There was an ulterior motive for Traci's wanting to stay in town. Between animal feedings, power tools, and lab construction, Traci was also an accomplished singer. With herring scales beneath her fingernails and the odor of dolphin breath in her nostrils, she took to the stage of Twin Lakes Church in its annual holiday concert. Work boots and bleach-stained sweatshirts were replaced with a black evening gown and pumps. Her long, dark brown ponytail was coifed and her lips painted. I'd never seen her in a dress, much less in makeup.

With a hundred other men, women, and children, she raised her voice in the pure joy of Christmas music. To those of us in attendance, the transformation was every bit as miraculous as the tidings of comfort and joy she sang about so beautifully.

The next morning Traci mucked out KP2's enclosure with a garden hose and scraper pole as usual. While she cleaned up discarded fish bits, there was no hint of what had taken place the previous evening. She was back to the earthly reality of KP2's stubborn refusal to eat.

The tough-as-nails trainer with the heavenly voice drew up one of the white Adirondack chairs from the truck transport. She had planned all along to share her holiday lunchtime with the seal. So in the warmth of the sealarium, she ate a turkey sandwich in silent thought.

KP2 watched from his pool, initially unsure of Traci's inactivity. She didn't move or try to force another fish to his lips. This was uncharacteristic of the trainer. Curiosity soon got the better of KP2 and he finally slipped onto the deck. The young monk seal inchwormed his wet body toward her, sniffing her pant leg in his usual greeting. When Traci didn't react, he nosed the cooler that contained his uneaten breakfast. His seated visitor continued to quietly eat her sandwich. Somehow the game rules had changed.

Slowly Traci opened the cooler and tossed a herring into KP2's pool. He watched as it sailed through the air and splash landed. Then the trainer tossed another and another. The dead fish shimmered and schooled as they sank. Immediately, the motion caught the seal's attention. In his mind's eye, the fish had suddenly come alive. KP2 whirled around and dove enthusiastically into his pool, gulping any herring along his swimming path. He circled the bottom perimeter of the pool, feeling for the fish with the long whiskers of his muzzle extended forward like a cat. He sucked larger fish into his mouth whole, then thrashed them when he surfaced as if they were still alive. As with the octopuses in Kaneohe, he tossed his fishy prey, pounding them into submission by shaking his head back and forth. Within minutes, the seal that had been reluctant to let one fish pass through his pursed lips had filled his stomach with over two pounds of herring.

With a full belly, KP2 once again slid onto the sealarium deck next to Traci. This time, rather than head to the heat lamp to sunbathe or nestle in a pile of toys, he crawled under the trainer's Adirondack chair. Doglike, the seal stayed curled at her feet, content in the simple human contact and sounds of the Pacific Ocean in his tiny ears. Traci just smiled knowingly.

Since the beginning, all the little seal really wanted was human comfort. He showed us that he was willing to sacrifice everything, even a square meal, for it.

BACK IN HAWAII, Christmas Day was heralded, as it was two millennia before, by the arrival of a mother. Early in the morning, RK22, the seal mother who had so abruptly abandoned KP2, returned to the location of his birth. RK22 didn't call for him or make any other sound. Yet she stayed on the Kauai beach for several days, rolling in the warm sand that had once been mingled with their blood. She nuzzled the ground and sniffed the air with no other seals in sight. She watched and she waited. Then as mysteriously as she had appeared, she slipped back into the water.

11.

Maui Wauwie

⸺

For all its accomplishments in the sciences, engineering, and the arts, UCSC, to the embarrassment of University of California administrators, has never gotten past its 1960s roots and geographical location. Named by *Rolling Stone* magazine as "the most stoned campus on earth," there is a perceived countercultural aura that attracts freethinkers, and the occasional free-smoker, to the school. Whether the reputation is deserved or not, pot farms indeed dot the surrounding redwood forests and can provide a seemingly endless supply of weed to students, medicinal marijuana establishments, and downtown pipe shops. And priorities are priorities for the campus. While budget cuts forced the library to cancel subscriptions to many scientific journals, preparations were still under way for construction of "Dead Central," a new library wing dedicated to the

Grateful Dead archives. I love our school, whether it is stoned or not, and given the local culture, I have become adept at differentiating the straight from the baked, ripped, floatin', chonged, and wasted. Consequently, it was not difficult for me to recognize that I had a stoned seal on my hands.

On the January afternoon in question, KP2 swayed and fixed his large, cloudy brown eyes on me. He stared without comprehension. If his lips had been capable of curling into a smile, I'm sure they would have done so. The smallest and most mundane objects in his sealarium—boxes, hoses, plastic floats—momentarily fascinated him. Then suddenly each was summarily dismissed as something new grabbed his attention. Back and forth the seal weaved across his enclosure. Every corner was explored; every detail inhaled with a snuff of his nose. Even his tiny pinhole ears seemed to droop on either side of his head.

"He's totally high!" Traci noted the obvious, shaking her own head as KP2 tried to wedge himself into the small space beneath the Adirondack chair she was sitting in.

There was little doubt that our endangered seal was "under the influence," although not by his own doing. After sixty days in his isolated, heated pool, separated from the rest of the animals at the lab, KP2 was finally ready for his last health exam before release from quarantine. One last roadblock prevented his move: we needed a blood sample to ensure that he was free of any diseases capable of infecting the seals, sea otters, and dolphins on the other side of the fence.

But the seal was not cooperating.

After so much handling during the previous twenty months of his young life—handling that included a blood sample on each capture, release, and transfer—KP2 was sensitized to needles. He was especially sensitized to people that came bearing needles. The moment a team of humans arrived in his enclosure trying to act nonchalant, KP2 knew that something was up. His sight might have been failing, but he could hear the crinkling from the sterilized paper wrapper of a syringe a mile away.

His previous blood samples had been drawn from the giant epidural vein that ran down his spine. The site was the seal equivalent of the crook-of-the-elbow needle stick for humans. I had taken blood samples from Weddell seals, harbor seals, and giant elephant seals from this "sweet spot." Compared with humans or sea lions, getting a blood sample from a phocid seal is remarkably easy. With the seal lying belly down on the ground, you palpate the hips, draw a triangle from the hips to the spine, and stick in the shallow depression formed by your thumb. It's the easiest vein in the world to hit. The trick is getting the seal to lie quietly while you poke.

KP2 knew the palpation sites and would have none of it.

"Watch this," Beau instructed me. With the seal lying on the ground next to him in perfect blood-sampling position, Beau began petting the quiescent animal. He started to run his hand down the seal's back. The seal never moved a muscle; instead, he seemed to enjoy the free scratches and attention. KP2 looked like a veterinarian's dream. Starting with the seal's head, neck, and shoulders, Beau petted and slowly slid his hand farther down. As soon as his hand reached the halfway point on the seal's back, KP2 immediately flipped over. The seal rolled completely onto his back, exposing his white belly. He spread out his front flippers, arched his neck, and opened his mouth wide in a "What the heck?!" posture.

No amount of encouragement could get the seal to turn back over. His flipping behavior was so ingrained that no one could touch

KP2's lower back, tail, or even hind flippers without the seal rolling over suspiciously. He never bit, but we also didn't push the sensitive animal to that point.

"Well, what are we going to do?" I asked as the seal eyed me warily.

"It's going to take time to break him of the habit," Beau remarked. "He's just like Puka. He's learned how to fight off the vampires."

Puka, one of the ex–Navy bottlenose dolphins waiting to meet KP2 in a neighboring pool, was renowned for his battles with needles and veterinarians. During his military stint in the Dolphin Systems program, discipline and schedules were essential. There was no time for wusses afraid of needles when soldiers' lives were at stake. To avoid having to sedate the dolphin for a blood sample, the Navy would gather five or six of the largest marine mammal trainers and soldiers to steady the animal as he floated in shallow water. The mere presence of a person carrying a needle, including myself on occasion, immediately transformed Puka from imperturbable dolphin into determined fighter. I was always astounded at how quickly and how far the dolphin could hurl grown men into the air with a flick of his tail. Often the inexperienced ended up with bloody noses and bruised ribs in the process of trying to get a simple blood sample out of the dolphin.

It was not the needles per se that Puka resented; it was the physical restraint that accompanied the procedure.

These days, after years of training, Puka quietly rolls upside down in the water and flops his large flukes into the waiting lap of a trainer sitting on the pool deck for his monthly tail vein blood sample. Despite his current passivity, he still harbors the memories of the old days and first cocks his big head suspiciously whenever I walk by his pool. But after checking my hands for a syringe, he rolls over in the water for a friendly body rub. Sometimes it is all about how you ask.

. . .

KP2's VETERINARIANS in Hawaii had used a combination of drugs and manual restraint to control him for his previous medical exams and blood draws. This worked well when he was a small pup of thirty pounds, but the method was becoming less effective every day. Now that he was back to eating, KP2 was growing so rapidly that physically holding him down for a blood sample was not an option. In time he would weigh over four hundred pounds and be able to shake off a team of veterinarians with needles as easily as Puka.

Eventually, Traci and Beau would train him to voluntarily participate in the blood-testing sessions like the recalcitrant dolphin and the rest of the animals. Veterinary procedures were a practical part of their behavioral repertoires, just as they would be for any well-mannered dog or cat going in for a checkup. The only difference is that these animals have the advantage in size, strength, and water.

KP2 was certainly smart enough to learn the behaviors. There just wasn't enough time to overcome his fear of needles before we needed to move him out of quarantine.

After consulting with Dr. Gregg Levine, KP2's Hawaiian vet, Dr. Casper settled on a light dose of Midazolam, an ultra-short-acting muscle relaxant. The plan was for KP2 to quickly fall asleep, allowing us to obtain a blood sample before he had a chance to even think about flipping over. What we hadn't realized was that Midazolam, along with its anxiety-reducing properties, was a hypnotic. Rather than falling asleep, the monk seal went tripping with the best of the Santa Cruz "weed-whackers."

As we waited for the drug to relax the seal, he bumped unsteadily into walls and people. All objects were sniffed and subsequently bitten. KP2 was not a mouthy animal, but under the influence of Midazolam he used his enormous whiskers and mouth as a means of exploring. Wandering over to me, he looked up with a hazy expression, tickled my leg with his probing whiskers, and then slowly put his mouth around the toe of my shoe.

"No-ooo," I warned the fuzzy-thinking seal. He then rolled over to nip at the sandals of a student volunteer, and nuzzled and chomped on a garden hose, nearly puncturing it with his sharp teeth. He climbed and slipped down walls in sluglike slow motion.

"He's not very respectful," Traci remarked when KP2 tried to slip his wet, heavy body across her foot, up her leg, and then into her lap.

"That's because he sees three of you right now." Beau laughed as he tried to guide KP2's bulky body off Traci and then out of a corner.

Resisting any helping hands, the little seal finally rolled onto his back and shut his eyes in perfect repose. He mouthed at a plastic float by his head and finally lay quietly with his belly to the sun.

He wasn't asleep; he just didn't care.

Taking advantage of the seal's brief intermission, four of us righted KP2 and made short order of drawing his blood sample. It was the usual phocid easy stick and the light-headed seal couldn't have cared less.

Afterward KP2 headed to his pool for a snack, albeit in slow motion—slow even for a seal. Leisurely slipping into the water, KP2 waited expectantly by the edge of the pool as Traci retrieved his food bucket. Outwardly, he seemed to have recovered from the effects of the drugs. My impression quickly changed when Traci offered him a thawed fish head. The relaxed seal sunk down to the bottom of the pool and proceeded to nudge the dead fish through the water in an effort to make it swim. He mouthed the fish head, spit it out, and then nudged. He pushed it along with his nose and held it in his mouth, only to let it roll off his tongue. Across the pool bottom he scooted the fish head until Traci gave up trying to feed him.

"Obviously, Midazolam is not all that 'fast acting' in seals," she quipped as she dumped out the rest of KP2's lunch. It would take a long afternoon snooze in the sun before the Hawaiian monk seal returned from his trip to the "higher" side of Santa Cruz.

ONE WEEK AFTER KP2's blood test, I found myself in a dental surgeon's chair dealing with a tooth that had cracked under the intense cold of the Antarctic. As I looked at the IV drip feeding into my arm, I asked the surgeon what was in the line.

"Midazolam," he replied. "You're not going to care which tooth we are working on!"

I started to sweat, concerned that I would start crawling up the walls or biting the surgeon's shoes. Apparently, I incessantly repeated this concern and something about Pele's revenge to the surgical team during the entire procedure, although I have no recollection of any of it.

There was a certain level of scientific curiosity and egalitarianism in having to experience what KP2 must have felt. My teacher side decided that I should share my newfound enlightenment with Beau, Traci, and my graduate students. The moment I got home I sent the following e-mail:

So yo know all ofths drugs we give to msrine mamals- vlaiu, ketamine medazolam, etc. Eveyt wonder what theyt are

thingking? We;; this stuff maked you totally WHACKED?!?
Whoooo baby,

Think I better for to sleep.

now get to werk—kreshaeeee- the whip cracks!Q??1

Terei

No wonder the seal was trying to make a dead fish swim.

12.

The Lost Seals

KP2 inchwormed his way across the cement walkway of the marine lab, following Beau like a faithful dog. The trainer found nothing unusual in this as he was used to traveling with a Hawaiian dog by his side. Kali, the short-haired puppy he'd adopted in Hawaii, had been an inter-island frequent flyer when Beau commuted to dolphin facilities on Oahu and the Big Island. Flight attendants knew Kali by name as well as her penchant for mangos, papaya, and *lilikoi* (passion fruit). As a puppy Kali had been allowed to play in the plane aisles, to the amusement of the other passengers. Through the use of peanut butter rewards, she had learned to walk politely next to Beau, ignoring any distracting sights and sounds around her. Now the "poi dog" was a mainlander and Beau's constant companion. She was also a test dog for training techniques that Beau would design for KP2. The only difference was herring tails replaced peanut butter for the seal.

"Come on, Mr. Hoa," Beau called, using the seal's lab nickname. KP2 responded excitedly by bouncing out of his small quarantine enclosure and onto the sidewalk. For the eighth time in his young life KP2 was headed for a new home. This time there were no U.S. Coast Guard or military flights, no vans, trucks, or police escorts. Instead, with a snort and a chuff the small seal humped along behind Beau, slowly making his way toward his big new pool, leaving a snail trail of water behind him. For the first time since his arrival on the mainland he was free to explore. The gate to the main compound was swung open, suddenly revealing the colors of the California coastline to the wide-eyed seal.

The sights and sounds of people and animals, the expanse of blue sky, the oceanic vista as well as the exercise initially overwhelmed the young seal. Periodically he stopped to look around and take it all in. With his rolling, stumbling gait and head raised in curiosity, he reminded me of the mainland dogs brought by their owners to Kaneohe beaches immediately after release from the state-run quarantine facility. Dogs imported to the islands were kenneled by the Hawaii Department of Agriculture in a four-month quarantine. Upon release, the previously jailed animals could only stare blankly at the roaring surf on their first trip to the beach, seemingly wishing for the four walls they'd considered their Hawaiian home.

KP2, on the other hand, barreled through the gate wanting to investigate every nook and cranny of the marine lab following his isolation. Since his arrival in Santa Cruz, he had listened to the calls, screeches, squeaks, laughs, and whistles that were the pulse of the lab. Finally, he was able to connect faces to the sounds.

The most vocal and certainly the loudest residents sat on a large outdoor wooden perch across the compound. A collection of ragtag orphaned parrots looked down their beaks at the newest resident humping his way toward them. Most days they whittled away at their perches, sunned themselves, and literally danced for human attention. KP2's inchworming progress was something new. Wiki, a large, faded pink Moluccan cockatoo, watched with a sideways turn of his head. He ruled the flock that included Junior, an umbrella cockatoo, and Mary, a petite sulfur-crested cockatoo that incessantly repeated "Mary, Mary, Mary . . ." in a hoarse chant. The three birds as well as a handful of smaller parakeets had been deposited on my lab's doorstep by owners that could no longer handle them. Some of the birds bit, one had developed the nervous habit of pulling out its own feathers, and all were neurotic in their own charming ways.

The birds had a vocal repertoire that went well beyond the requisite "hello." Wiki tried a low catcall whistle on KP2 to no response. Junior followed with a series of bird chuckles. He was adept at mimicking the laughing styles of the people working at the lab. The bird had Heather's giggle, Andy's chortle, Amber's titter, and Maria's hearty guffaw down with such disarming accuracy that we were often left looking for the person following his impersonation.

Junior was also partial to women, bobbing and dancing for females who walked by in a mesmerizing display of head feathers after he laughed seductively. Men he just bit. He quickly grew bored with the fumbling seal and went back to cruising for women with his displays.

In the large pool next to the birds swam the two quietest but biggest personalities of the lab, Puka and Primo, the U.S. Navy dolphins I'd worked with in Kaneohe. Both had failed to rise to the rank of mine hunter. The more romantic visitors to the lab said that the dolphin's instinctive pacifism was behind the smiling animals' refusal to join in the business of man's destructiveness. In truth, these two dolphins were just obstinate. Primo, the smaller of the two (and I always suspected the smarter), was brilliant at locating mines. However, he worked only when the mood struck him. Any nearby activity, including schools of swimming fish or a tour boat cruising in the distance, easily distracted him and he'd take off to play. Puka, the pretty boy with dashing gray racing stripes on his melon, was the lover. He had no desire to venture into deep water and find submerged mines. Female dolphins were his target and eventually led to his dismissal from the Dolphin Systems. In the end the Navy could not risk the lives of men and women with such unreliable oceanic watchdogs.

With the 1994 closure of the Navy's Dolphin Systems program in Hawaii, Puka and Primo were transferred to California along with Austin and me. When the Navy decided to officially retire the boys, I offered them a home at Long Marine Lab. Since then they had spent their days in the company of an enthusiastic corps of volunteers consisting primarily of young female students from the university. The women unabashedly doted on the two dolphins, which fit Primo's and Puka's personalities perfectly.

The remaining three occupants of my lab, Wick, Morgan, and Taylor, were also dropouts, this time from the Monterey Bay Aquarium sea otter rehabilitation program. Like KP2, the three male otters had been found injured or abandoned by their mothers during coastal storms. The aquarium had raised them from pups to juveniles and released them back into the bay. One by one they returned to the aquarium because they failed to recognize that they were sea otters. In the wild, the otter boys climbed on sea kayaks, harassed divers, and generally roved around the coastal waters like gang members. But it was

Morgan who earned a reputation that still had reporters coming to my lab to photograph him.

Morgan the sea otter was publicly known as a convicted rapist and killer; in truth he was little different from other wild sea otters. When released by the aquarium, the large male otter had swum north to an inland waterway called Elkhorn Slough. Cruising along the shoreline, he encountered another ocean creature of the same approximate size and shape as himself. These new creatures sunbathed along the shore, gathered in groups, and paddled in the shallow waters of the slough. Morgan was infatuated with them. Using his strong forearms, he selected a mate as any healthy male sea otter might. Unfortunately, the object of his affection was a harbor seal pup. In the process of coupling the seal pup drowned. Undaunted, Morgan chose another and then another, with the same dreadful consequences. For months Morgan terrorized the harbor seal colony of Elkhorn Slough until he could be recaptured and "sent to college" at my lab.

Over the years, under Traci's tutelage, Morgan was rehabilitated. He was turned into one of the best-trained and most valuable research sea otters in the country. Morgan could dive and swim on command. With Traci holding on to his front paw, he would rest quietly on the water surface while researchers took heart rate and temperature measurements. During his veterinary checkups he would crawl into a kennel when requested and then lie still while a blood sample was taken. He was unflappable and reliable. But we never dared put him in the same pool with another animal. There was no sense in tempting fate.

As KP2 MADE THE ROUNDS, I saw my lab for what it was: a collection of outcasts. All of the animals in one way or another had failed in their respective societies. As such they were deemed by others as nature's surplus. Many would have faced certain euthanasia had I not intervened and offered them a last chance. I saw something different: rather than misfits, they were species ambassadors.

Over the years the dolphins had been employed as models for killer whale metabolism and involved in projects to determine the effects of oceanic noise on marine mammals. The otters were used to predict the unique caloric, metabolic, and thermal requirements of their wild counterparts along the California coast. Even the birds had a job: they taught the volunteer students in my lab the nuances of operant conditioning in animal training. We could not afford training mistakes during research with the marine mammals. So the abandoned birds were enlisted and proved to be the perfect, enthusiastic training ground. With the help of Beau and Traci, every animal in my lab was transformed into an invaluable partner in our conservation efforts. Fittingly, the whole place was directed by someone whose greatest ambition in early life was to be a dog.

"THIS IS IT! Come on up! Water!" Beau encouraged the monk seal to climb up a handicap ramp leading to his new home. KP2 slowly inched his way onto a deck next to a saltwater pool thirty feet in diameter and eleven feet deep. Traci and Nate Moore, a facilities technician for the lab, had created a giant version of the previous sealarium by walling the entire area with Plexiglas and topping it with the greenhouse plastic sheeting.

KP2 lost no time wriggling into the heated 85°F water. Unbeknownst to him, there was a small underwater viewing window. On his first dive, he suddenly saw our smiling faces in the glass. He somersaulted several times in front of us and then sped away, enjoying the freedom to swim and dive again.

I had been warned from animal caretakers at SeaWorld that monk seals were exceptionally curious and would stick their heads into any opening. All drains had to be covered to prevent an accidental drowning. KP2 lived up to his species's quirkiness by immediately becoming fascinated with the water outflow pipe. Shoving his muzzle into the pipe, he let the water stream past his whiskers for minutes on end. The water came directly from Monterey Bay, and I wondered what the seal found so attractive in it.

"Maybe he can smell or taste wild seals in the water," I suggested to Traci and Beau as KP2 floated blissfully with his eyes closed under the waterfall.

To divert his attention from the pipe, the trainers provided KP2 with a wide variety of toys, including plastic boat bumpers, barrels, and floats. Two items quickly became his favorites, both reminders of his days in Hawaii. The first was a bowling ball. Like the coral rocks he'd maneuvered with his muzzle in Kaneohe Bay to find fish, KP2 head-butted the heavy ball around and around the bottom of his pool until the floor paint began to chip. He wasn't looking for fish, he just enjoyed underwater games.

KP2 latched onto another toy with such fervor that we couldn't wrench it away from the seal's flippers. It was a pink plastic children's slide that floated on the water. Instantly, KP2 began surfing and sleeping on it. In the water, he flopped his body on top of the board with enough force to create a wave for surfing. Then he extended the ride by propelling with his eggbeater hind flippers. On deck the pink slide became his tanning bed.

The young monk seal obviously recognized the pink float from his puppy days when his caretakers had conducted swimming lessons with him in the pool at the Kewalo Research Facility. It also reminded him of the pink boogie board from his wild wharf days, surfing with the children of Molokai in the warm shallow waters of Kaunakakai. Pink was the color of fun.

WITH KP2 OUT OF QUARANTINE and finally settled into his home in Santa Cruz, the mystery surrounding him immediately dissipated. Journalists stopped calling, special visitor requests declined, and a new, welcome equilibrium in my lab took hold. Finally I could focus on the science of monk seals. The pressure was on from Washington to prove that transferring the monk seal to the mainland had been worth the trouble. Somehow we had to transform this rambunctious animal into a research partner that would enable us to decipher the unique biological needs of his tropical seal species.

"It's sad, isn't it?" I said to Beau and Traci as we watched KP2 frolic in his new pool. "We've got less than fifty years to figure out how to save the monk seal before the species is gone."

My scientific career had begun with one monk seal extinction and appeared to be on the verge of ending with another. KP2's cousin species, the Caribbean monk seal, was already gone. These beautiful dark brown seals with cream-colored bellies, once called "sea-wolves" by Christopher Columbus, were killed and eaten by the Italian explorer's men more than six hundred years ago. The species had disappeared from the warm Caribbean waters during the beginning of my scientific journey. While I studied for my PhD, the only pinniped to call the Gulf of Mexico, Haiti, and Jamaica home never showed its nose. After waiting thirty years the National Oceanic and Atmospheric Administration's Fisheries Service officially declared the

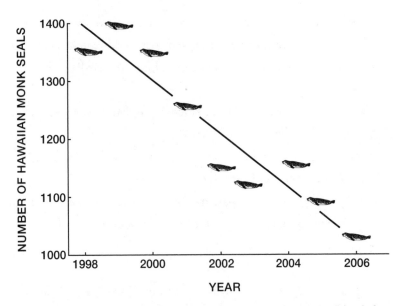

Decline in the number of monk seals in the Northwestern Hawaiian Islands from 1998 to 2006. Each seal symbol represents the total count for that year. The black line indicates a 3.9 percent annual decline in this subpopulation.

NATIONAL MARINE FISHERIES SERVICE, 2007

Caribbean monk seal extinct in 2008. I never even looked up from my books.

Now Hawaiian monk seals were teetering on the precipice of extinction. Reasons for the dismal state of the species spanned the International Union for Conservation of Nature and Natural Resources (IUCN) list of criteria for declaring a population endangered. Marine pollution, vulnerability to disease, rapid changes in habitat quality, competition with fishermen, shark attacks, and infighting among males and females were all factors.

Regardless of the cause, the population trajectory was clear. If the population trend was left unaltered, our children's children would know the Hawaiian monk seal only as a picture in a book. In the wake of the seal's disappearance, a generation of baby boomers who had given birth to the environmental movement would be left to explain how we let this unique creature slip through our fingers like a handful of sand.

"Well, that's the whole point," Traci responded matter-of-factly to my thoughts. "We'd better get to work!"

In her first small step toward saving KP2's species, Traci walked deliberately into the food prep kitchen to wash the monk seal's fish bucket. With her abrupt departure, the little seal popped his head curiously over the edge of the pool wondering what was coming next.

WHILE THE MEMBERS of my team continued to have nightly stress dreams over the responsibility of caring for such a highly endangered species, outsiders walked right by our monk seal. To others, KP2 was a small, nondescript gray seal, albeit from Hawaii, who enjoyed sunning himself. Like the small-mammal exhibits quickly passed over at zoos, the seal was summarily ignored by visitors who were quickly caught up in the antics of the dancing birds and the constant smile of the dolphins. We learned that, as in Hollywood, when it comes to being endangered, looks are everything.

Unfortunately, in a world of limited resources this meant that an animal's external appearance rather than ecological significance or population status dictated which species lived and which died. For many endangered animals, survival was cast in black and white. Frank Todd, a retired curator of birds at SeaWorld in San Diego, once remarked, "If it were up to me I'd have a zoo filled with just black-and-white animals." As the man responsible for bringing black-and-white Antarctic penguins to an aquarium that had made its reputation on black-and-white orcas, there was a lot behind his statement. Over the years Frank watched the public fall in love with, and pay amazing sums of money to see, black-and-white animals.

He imagined a zoo filled with killer whales, panda bears, and penguins. "You could make millions!" he said, laughing, which is precisely what he did for SeaWorld.

Frank had stumbled onto something that scientists had already discovered. There is a deep-rooted human attraction to black-and-white animals with disproportionately large heads and large, dark eyes. Biologists refer to it as the "cute response." The epitome of the phenomenon was the appeal of Mickey Mouse.

For good and for bad, the cute response has had a powerful influence on the conservation movement and how conservation groups appeal to the public for donations. It does not take much imagination to morph the beloved Mickey into fluffy white dome-headed harp seal pups or the iconic symbol of the World Wildlife Fund, the panda bear. These attractive symbols for conservation move people.

Humans have an innate, biological drive to nurture and protect living beings with babylike qualities. Rounded tummies, big heads and eyes, and stunted limbs bring out the parent in us. Even as a scientist who recognizes the underlying biological mechanism, I am not impervious; hence my attraction to neotenic corgi dogs and seals. Add in the cute factor of black-and-white coloration and humans quickly fall victim to the instinctual desire to nurture. When it involves wild animals, we end up with conservation by the heart.

Instinctual, heartfelt conservation drivers have instigated public outrage about (and Paul McCartney visits to) the Canadian harp seal hunt in which rotund, teary-eyed white pups are killed as they sleep on pristine snow. By comparison, little has been said concerning the northern fur seal harvest in which needle-toothed brown otariids are slain while curled in the Pribilof Islands' mud. Conservation by the brain would dictate a reversal in attention. The IUCN, the watch-dog group for identifying endangered species, classifies the harp seal as a species of "least concern." This is due to an increasing population that currently exceeds eight million harp seals across the Arctic. In contrast, the northern fur seal is slipping toward endangered status as its population suffers through a devastating 6 percent decline each year even in the absence of large, commercial hunting operations.

This is not to say that we shouldn't protect all animals. The question is, where should we begin?

The Hawaiian monk seal, due to its isolation in the middle of the Pacific Ocean and its unpretentious gray fur, seems to melt into the endangered species background. It is difficult to care about an endangered animal whose greatest talent seems to be sunbathing on beaches 1,860 ocean miles away from everywhere. However, the Hawaiian monk seal, with a population of only eleven hundred surviving individuals, has reached the pinnacle of IUCN listing categories: "critically endangered." There are fewer Hawaiian monk seals than panda bears or the newest icon for global warming, the polar bear. If the entire world's population of Hawaiian monk seals were lined up nose to tail, they would not even span the length of the Golden Gate Bridge in San Francisco. There are more minutes in a day and more jewels in Princess Diana's wedding tiara than there are Hawaiian monk seals. Which should humans consider the more precious?

"WHAT ARE WE GOING to do with you?" I asked as the seal soaked my shoes by placing a wet chin on my foot. I resisted the urge to pet his wet round head. Our official mandate from the National Marine Fisheries Service was to care for the nearly blind seal until a date for his cataract surgery was set. But his species needed so much more. The seal seemed to sense the concern in my body language and cuddled up even closer. I ignored his overtures as I tried to decide my next steps.

The Hawaiian Monk Seal Recovery Plan, NOAA's handbook for saving the species, was 150 pages long. Twenty-one pages presented the intimate details of the population decline over the past half century; less than three pages were dedicated to the inner biology of the animal. The meager section began with the statement "Comparatively little physiological research has been done on Hawaiian monk seals."

The reason for the paucity of data was obvious. Wild Hawaiian monk seals were too few in number, the population was too fragile, and the species far too endangered to handle for biological studies. Like so many other endangered animals, the Hawaiian monk seal had fallen into a bureaucratic trap. The same governmental regulations that protected endangered animals from human disturbance also prevented scientists from touching them. Consequently, Hawaiian monk seals and the other hands-off endangered species were in danger of being counted right into extinction.

KP2's mission was clear to me. He *was* different. Abandoned, handicapped, and in captivity, KP2 was not under the same governmental restrictions as his wild family. The troublemaker seal with a fondness for people was the one Hawaiian monk seal we could study in detail.

"We're going to fill the missing pages of the NOAA Hawaiian Monk Seal Recovery Plan with your biological data!" I began pacing the deck, causing the wet seal to unwind from around my legs.

KP2 retreated into his pool. I watched the small seal submerge and then resurface wearing a dog toy ring on his muzzle. He proceeded to shove the rubber dog toy into the outflow of the water pipe. He pushed it in only to have the toy spring back and sink. Then he retrieved the ring, positioned it on his muzzle, and started the process over, repeating the game again and again.

I doubt there could have been a more unlikely candidate for saving an entire species than KP2.

PART III

Survival

13.

Instinct and Intelligence

We had a lot to do in a short amount of time if KP2 was going to tell us about the biology of Hawaiian monk seal growth and development. The young seal was quickly gaining weight and growing broader by the day now that he was eating. Each day that we delayed was another data point lost. The transition from pup to juvenile had already occurred. I first noticed that he was no longer a puppy when the top of his head went from round to flat. He could easily balance toys on his blocky skull, while retaining his disarming phocid smile. At twenty-one months he was quickly entering young adulthood, although age did nothing to diminish KP2's playfulness.

I wanted to document how the young seal made the life transition from seal adolescent to adult. More than 80 percent of Hawaiian monk seals died at this stage in the wild. I suspected that part of the problem had to do with the extraordinary number of calories these young animals needed for body growth. Just like humans, young growing sea mammals typically have enormous appetites, so I devised a study to monitor KP2's metabolism as he developed. All he had to do was rest quietly in the water while I measured the oxygen level in his exhalations each week.

To accomplish our goal, Traci and Beau created a crash course in animal training for KP2. Nearly overnight, the young seal had to learn two important concepts: how to sit still and how to be polite around people and expensive scientific instruments. Like any energetic two-year-old, the seal's first response was "No!"

CONSIDERING HIS SUNBATHING HABITS, I thought that KP2 would be adept at sitting still. He seemed to excel at doing nothing but lounging in his sealarium. That is, until someone entered his enclosure. Immediately, KP2's enthusiasm for human company overtook his monk seal reserve. No matter who arrived, he'd surf in a wave of watery excitement, fly onto the deck, cuddle up to shoes, crawl onto scientific instruments, and generally crowd the person's escape. As charming as this was at 160 pounds, I feared what would happen when he reached his full 400-pound adult size. He was bound to flatten someone or all of my instruments in his exuberance.

This was where Traci, with her calming touch, excelled as an animal trainer. She had miraculously transformed Wick, Morgan, and Taylor, the frantic sea otters of Monterey Bay, into cool, confident research assistants. To curb the otters' habit of nervous grooming and swimming in circles, she taught them to float with their hind flippers touching the side of the pool. By focusing their energy into constructive rather than destructive behaviors—a technique borrowed from teaching children with attention deficit hyperactivity disorder—she had their attention. Research tasks quickly followed thereafter. I trusted that she and Beau could do the same for the boisterous KP2.

Secretly, I was envious of the trainers' position and the intimate relationship they were developing with the seal. Each day they worked on nurturing the animal's trust by talking to him and providing daily body rubs. He would sit quietly as they medicated his eyes with a cooling salve. The seal responded to their attention with obvious enthusiasm.

Yet no matter how hard KP2 tried to win me over with the same buoyant greetings, I could not bring myself to pet him. Somehow allowing myself that privilege felt like a betrayal to our science.

I also recognized my own limitations when it came to animal training. I possessed the best and the worst qualities of a trainer. On the one hand, I could read an animal with ease, an essential skill for any trainer. On the other, I was a pushover who believed in and even encouraged the wildness of creatures. Animals knew it and took advantage. Their behavior made me laugh and was most amusing at those times when they got the better of humans, especially if it was me. Some of my best science came from such animal antics.

My only formal schooling in animal training was short-lived and occurred when Austin was a ten-week-old puppy. Under the rustle of coconut palms that fenced a Kaneohe baseball field, we attended our first puppy obedience class. Self-taught in the natural rhythm of animal behavior, I balked at the rigidness of manipulating animal minds through a choke chain.

"You must show your dog who is alpha!" a big-bosomed instructor in a muumuu and wide-brimmed flowered hat demanded. "Otherwise they will *rule* you." To demonstrate, she tried to choke the spirit out of a border collie–Dalmatian mix to the high-pitched squeals of the dog and the horror of its nine-year-old owner.

"Acch! Whoever bred such a dog should be arrested!" the struggling instructor shouted, finally releasing the spotted dog with disgust.

I removed the chain from Austin's tiny neck. Off leash he was a star pupil following my every step through a gauntlet of older, panting dogs. We trusted each other implicitly; yanking on his neck was not necessary. The dogs in the class instinctively knew what to do. I felt that this particular teacher did not.

With the choke chain on, Austin and I suffocated on both ends of the leash. He'd lie down and then roll over on his back with his stubby corgi legs pumping the air. Instead of correcting him with a jerk of the chain, I laughed into crying. The instructor yelled. The dog kicked. And I ended up rolling on the grass with my puppy. We were the bad kids of the class who eventually flunked out. From then on the dog and I relied on our instinctual bond to get us through life.

That bond took us on high adventures all across the island of Oahu. On weekends we hiked the trails of Sacred Falls, and up to the top of Tantalus Mountain overlooking Honolulu. We camped on the North Shore sands to watch surfers challenge the pounding winter waves of the infamous Pipeline. On the beaches of Lanikai, Austin demonstrated the importance of instinct versus training by herding giggling children on the beach like sheep.

There was only one problem with my loose training style. It could not overcome Austin's irrational fear of water, a logistical challenge since we lived on an island.

With such short legs, Austin was a poor swimmer, making little forward progress in the water despite his best efforts. Recognizing his limitations only escalated his phobia into a paradoxical reaction whenever he encountered a pool of water. On kayaks he stared fixedly at the receding shoreline, and then jumped off in a frantic attempt to swim back to perceived safety. On walks, if we approached a pond, stream, or pool he would plunge in, desperately struggling to swim to the opposite shore when he could have easily walked around. Fear destroyed his ability to listen or think smartly around water. No amount of coaxing or swim lessons helped.

I clearly lacked the animal training skills that KP2 needed.

WHEN IT CAME TO KP2's formal education, I turned to Traci and Beau to serve as "translators" between me and the seal. My role was to decide the science that would be conducted and the methods that would be used. The two trainers broke the complicated mission into a string of manageable, fun tasks for KP2. Instead of forcing the seal into our scientific studies, we requested his assistance in saving his species through elegant training techniques and a bucketful of fish.

"There is still one issue with KP2," Traci reminded me. "He refuses to eat when he is out of the water."

Before we could begin any science, we had to solve this lingering problem. Like a dog rewarded with a bone for shaking paws, KP2 had to make the connection between the research tasks and fish rewards. Currently, the seal always ran away with the fish before the trainers could tell him, "Good boy."

Although Traci had overcome KP2's resistance to eating dead fish, the seal dined only underwater. Regardless of whether he was offered a fish on deck or in a pool, KP2 would take the morsel in his mouth, submerge, and then play. He pushed, nuzzled, and tossed the limp fish into the air until he was satisfied that it was sufficiently killed. Only then would he swallow. During these play bouts Beau and Traci were left on the deck tapping their feet with a fish bucket at their side waiting for the seal to come back for the next fish. Mealtimes were dragging into hours—a smart trick used by KP2 to prolong his time with his human friends but challenging to my scientific schedule.

For the moment, training the seal to do anything was impossible. He was like a dog that took a bone from your hand and then ran off. All science was stalled until the seal learned to eat on the water surface and on land, behaviors that obviously went against his wild marine mammal instincts.

Beau went back to basics and contacted as many people as he could find who had trained monk seals. His colleagues were somewhat less than encouraging, telling him that monk seals were not the easiest animals to work with. They called them "true seals," which was interpreted to mean independent, difficult, and generally snotty.

More than one trainer told him, "The monk seals' indifference is hardwired into their personalities. Good luck!"

Rather than becoming discouraged, Beau began a long, methodical study of KP2 to peel back the layers of the seal's psyche. The trainer rose early in the morning before the sun had appeared on the horizon, before anyone else had arrived at the lab, even before the cockatoos, sea otters, and dolphins had opened their eyes. The lab was dark, silent, and peaceful, a rare reprieve from the usual daytime chaos.

Quietly Beau entered the sealarium and simply sat on the deck by the seal. He watched every move of the animal and noted all the things KP2 found stimulating or interesting. He listened intently to the underlying rhythm of the marine lab, experiencing what the young seal experienced. He did this without uttering a word that would break the moment. The dripping of water, the low hum of pump filters, the call of seagulls, and the whoosh of pelicans soaring above the incoming tide materialized and sharpened in his silent reverie.

Soon Beau realized that the seal's diminished eyesight heightened the value of all his other sensory cues. Movements, sounds, touch, and smells painted an environmental picture for KP2 even if he couldn't see details. Through his whiskers, nose, ears, and skin he was able to connect with his surroundings. These were the things that mattered to the seal.

One morning Beau found KP2 lying on the bottom of the pool and his heart skipped a beat. Then he noticed that the seal's hind flippers were slowly sculling. KP2 had his head shoved onto a shelf that was formed by the corner of the pool gate and the wall. He seemed to be intensely working on something underwater, so the trainer sat back and watched.

KP2 had retrieved all of the toys the trainers had given him and was stacking them one by one on the gate shelf. Balls, deflated floats, and large dog chew toys were all piled in one place. KP2 busily stacked and restacked the items until they stayed put. His behavior was reminiscent of his Hawaii puppyhood when he used to pile coral rocks to catch small fish and crabs in his net enclosure in Kaneohe. But there were no fish to be found in the pool. It didn't make sense.

At last KP2 came up to take a breath of air, spraying water on the trainer's shorts as he loudly snorted to clear his nose. In the subdued light of dawn it was difficult for the seal to see Beau when he sat motionless. Instead, KP2 listened to the rise and fall of Beau's breathing and then felt his way with his long whiskers, mapping the location of the water pipes and the pool edge, and finally finding the seated trainer.

Beau now understood the purpose of KP2's toy stacking. The stack was a landmark. Just as visually impaired humans use the arrangement of the furniture in their homes as clues for navigation, the seal relied on the predictable placement of objects in his sealarium to move as freely as a sighted individual. By placing all his toys in a known location, the nearly blind seal could use his internal map and whiskers to always find them. The trainer thought, why not use the seal's natural whisker-assisted mapping to help him find food to eat on land?

To halt the seal's habit of flopping back into the water to eat each fish one by one, Beau decided to bring the water to the seal. During the next mealtime, the trainer called KP2 onto the deck with a quick "Come, Hoa." The seal popped out of the water as usual looking for a new adventure. Beau held a target pole, a two-foot-long stick with a yellow plastic float stuck on the end of it. KP2 knew the drill. Touch the ball with his nose and get a fish. His next move was to grab the fish in his mouth and roll backward into the water to play.

This time was different. This time the trainer raised the fish above the seal's head and let him feel it with his long whiskers. Next, in an unexpected move Beau threw KP2's fish into a dishpan filled with salt water that he had positioned on the deck between him and the seal. KP2 stared at the trainer, then at the pan of water with the fish lying on the bottom, and then again at Beau.

"Good!" Beau said sharply. In an instant the seal went for it. KP2 dove into the pan, nuzzled the fish, and swallowed. Although his body was dry, his head was wet and that was good enough for eating.

In time and with a series of approximations, Beau soon had KP2 sitting on deck eating from his hand. The trick was all in the whiskers. Rather than hand the fish to the seal as one would do for a dog, Beau held the fish above KP2's head. Only after the seal had explored the fish thoroughly with his whiskers would he open his mouth and allow the trainer to drop it down his waiting gullet.

KP2's REPERTOIRE OF BEHAVIORS increased exponentially from that point forward. He learned to sit still and roll over for veterinary exams. Even more impressive, he quietly allowed teams of ophthalmologists to monitor the progression of his cataracts by studying his eyes. This required stringing together a series of trained behaviors, beginning with hauling out on deck without soaking people, lightly placing his chin on a small plastic stand, and opening his eyes wide despite the shining lights of a camera or retinoscope. These behaviors were especially useful for the trainers as we needed to medicate KP2's injured eyes with drops every day. Remarkably, he never blinked. I wish I could say I was as good a patient during my eye exams.

"Watch this," Beau instructed me cryptically one day. KP2 was on the other side of his pool playing his buoyancy game with a coconut the trainer had given him, the same way he had in Kaneohe.

"Come, Hoa!" Beau called to the seal. KP2 immediately began surfing across the pool in his usual mounting tidal wave. I fully expected my shoes to be covered in salt water when the wet seal bounced onto the deck and then banged into my legs. But this time the surfing seal stopped short. Before he reached the edge of the pool he put on the brakes and slowly floated toward me. The overenthusiastic surfing seal who had once almost drowned Ingrid Toth in a friendly bear hug near Kaunakakai Wharf was now the picture of politeness in the water. KP2 had finally learned to respect the personal space of humans.

"He even knows how to pick up his toys," Beau said proudly. Following a loud "Retrieve" from Beau, KP2 swam through the pool collecting the coconut, balls, and floats. One by one he pushed each toy across the water with his nose and then deposited them into Beau's waiting hands. With his bad eyesight, the seal relied on Beau's verbal commands and his whiskers to accomplish the task.

"Remarkable!" I was duly impressed with the transformation in KP2. "Now on to the hard part—science!"

The most critical task for our upcoming metabolic experiments seemed simple but would challenge the seal's instinct in the water and his boundless sense of fun. He needed to float quietly. Furthermore, he had to hold his position for at least twenty minutes beneath a metabolic hood. During the test he would have to ignore the pulsing of a vacuum pumping air in and out of the hood at hundreds of liters per minute. He had to disregard Wiki, Junior, and Mary, who called incessantly, demanding to be the center of attention, as well as shut out the people and dolphins bustling around the lab. KP2 had to focus on what he was doing without making a sound, but he could not simply fall asleep. To make the task even more difficult, he needed to perform these behaviors every other day first thing in the morning on an empty stomach, with no fish reward until we'd completed all the measurements. These were the strict biological conditions that I needed in order to measure his metabolic rate. Deviating from any one of these conditions would make it impossible for me to interpret his data. Digestion, activity, growth, temperature, and the peculiarities of monk seal biology could affect KP2's metabolic rate. My job was to decipher how each of these factored into the daily caloric demands of the young monk seal.

I knew that I was asking a lot of KP2. Other scientists might have just shoved the young seal into a water-filled metabolic box and waited for the animal to eventually calm down. I preferred voluntary participation in my science to encourage a more natural physiological response.

"Do you think he is smart enough for this?" I queried Beau before he began training KP2 for the experiments.

"He has shown me that the only limitations he has are the ones that I place on him." There was no question in the trainer's mind.

I considered the words high praise coming from someone who was used to training dolphins, purportedly the most intelligent of marine mammals. Personally, I have always been uncomfortable when people ask me which animal I think is the most intelligent. To me all animals from dogs to dolphins to seals and sea otters seem intelligent in their own unique ways. Furthermore, I find it impossible to define intelligence without an IQ test specific for each species. What would be considered a mark of intelligence for humans is not necessarily valuable for the survival of an animal in the wild, just as standard tests designed for the general population could not predict the street smarts of a gang member.

Admittedly, dolphins have proportionately larger brains when compared to other mammals. People have naturally linked this morphological characteristic to intelligence. Based purely on weight, the bottlenose dolphin brain is one of the biggest, averaging 1,600 grams (3.5 pounds). By comparison, adult human brains are slightly smaller, at 1,350 grams (3.0 pounds). If big brains equals smart, then dolphins clearly trump people.

A fairer comparison, however, is to account for the differences in body size accompanying those brains. Here we use the encephalization quotient (EQ), a measure of relative brain size determined from the ratio between actual brain mass and predicted brain mass based on the size of the animal. Using these calculations, we see a slightly different picture. The EQ of bottlenose dolphins is 4.1—that is, their brains are more than four times the size that would have been predicted based on body size alone. This compares to 2.6 for the killer whale and 2.5 for chimpanzees. Dogs, cats, and seals have EQs that range from 1.0 to 1.4. By far the champion is the human, with an average EQ of 7.0. Thus, pound for pound the dolphin is second only to humans when it comes to braininess, with the seal much farther down the list.

Determining how much of the big dolphin brain is actually relegated to computing power is a different problem. A bigger computer doesn't necessarily mean faster or more intelligent processing; that depends on internal chips and wiring. For dolphins a large portion of their brain wiring is dedicated to processing sonar information. This unique sensory capability would be akin to humans having X-ray vision. Brain power is certainly needed for this task, although science has offered no clear answers as to what proportion of the brain is involved.

It took a Navy experiment in Kaneohe Bay designed to measure Primo's brawn, not his brain, that changed my perspective regarding the intelligence of dolphins. The research task was an evaluation of the dolphin's heart rate during swimming. The Navy needed a report on the exercise capabilities of their aquatic watchdogs before they could deploy them on long-distance mine-hunting missions. I decided to use the same exercise testing methods developed for Olympic athletes, with a few modifications for dolphins.

For months my research team trained Primo to swim next to a moving boat, chasing it the way a dog chases a car. Our racecourse was the deep channel of Kaneohe Bay. With a boat driver, a trainer with a bucketful of fish, and Primo at our side, my crew and I spent our days cruising the blue waters training the dolphin for his physical fitness test. Often we discovered schools of brightly colored tropical fish along the way, and passed boats filled with tourists who screamed in delight. These excursions were truly the most delightful days for the dolphin and me.

On a beautiful clear Hawaiian day with calm, flat seas, we were finally ready for Primo's big test. To measure his heart rate, I had bought a $3,000 instrument that was so new to science that it was not even available on the market. The instrument consisted of a tiny microprocessor that fit inside a waterproof metal tube measuring six inches long and one inch in diameter. Two wires leading from one end of the tube connected to suction cups. When the suction cups were positioned on the skin on either side of the dolphin's chest, the instrument was able to record the electrocardiograph signals from his heart. The tiny high-tech instrument worked in the same manner as the beeping machine in an emergency ward monitors a patient's cardiac rhythm. This time the technology was going for a swim in the open ocean to record every beat of Primo's heart.

With a roar of the outboard motors we took off for the racecourse, giving Primo a short warm-up sprint before his official test. Once the dolphin and our boat had reached the deepwater channel, we stopped to outfit Primo with his heart rate monitor. Primo had practiced wearing the suction cups and the nylon harness carrying the microprocessor in his home pen. He was a pro and floated easily next to the skiff as we positioned the two body straps around his midsection and shoulders and pushed the suction cups onto his skin.

With a toss of a fish into his waiting smiling mouth, we were off.

"We'll start with a slow five-minute cruise at four knots," I instructed the boat driver. This was an easy walk for the dolphin, who swam smoothly next to our boat. I made notes, pleased with the session.

"Okay, now six knots for the next five minutes." The boat driver put the boat into cruise mode. Again Primo stayed by our side, fluking and breathing steadily as he swam. There was little doubt that he was working to keep up at this rate, which was equivalent to a moderate jog for the dolphin. I also noticed that turbulent bubbles were beginning to form around the harness straps. This was new and I hoped that the straps wouldn't fail. The dolphin's body was so well streamlined that we never saw turbulent wakes when he swam even at high speeds. The straps were clearly changing Primo's hydrodynamic advantage.

Regardless, the dolphin was holding steady, so I upped the speed to the equivalent of a dolphin run. "Eight knots!" I screamed over the roar of the outboard motors. This time the dolphin arched his dorsal fin out of the water to catch us and tried to ride on our bow wave. This was akin to cheating on his fitness test.

"Oh, no. No free rides!" I corrected the dolphin and he repositioned himself in the open water. Bubbles were now streaming off the straining harness as Primo swam. The water pressure was so great that one of the straps had twisted, but the suction cups were still in place. At one point, Primo swam up next to the boat, spiraled, and gave me a sideways glance. I never thought of dolphins as being able to change their facial expression. But if I could have interpreted his glance, it unmistakably said, "I'm getting tired of this." Despite the look, Primo kept swimming in proper position.

Encouraged by the dolphin's performance and the heart rate monitor, I shouted, "Let's try a sprint! Nine knots!" This time Primo leaped high, clearing the water completely with the harness and monitor wires dragging a wave of water with him.

All of a sudden the dolphin disappeared behind us. As I turned around to find him, our outboard motor coughed as the propellers jammed. We had caught onto something and I was horrified.

"Stop! Stop the boat!" I yelled. "We've hit Primo!"

We surfed on our own diminishing wake, coming to a bobbing halt in the water with no dolphin in sight. I frantically looked for any sign of Primo or blood. The pit of my stomach cramped. Water slapped the gunnels as we used binoculars to scan the horizon looking for the spray of a dolphin breath. The trainer tried banging Primo's food bucket on the side of the boat. No response.

Several nerve-racking minutes passed, and I cursed myself for asking so much from the dolphin. Then slowly a gray head emerged at the stern of the boat.

"Primo!" I shouted in relief. The dolphin came alongside our boat carrying something in his mouth. Immediately I realized that it was his heart rate harness. He spit the bundle of nylon straps, wires, metal, and a suction cup into his trainer's hands.

"What the heck, Primo?" The trainer handed the tangled mess to me, apologizing. "I'm sorry. Looks like we're finished for today." He did a quick inspection of the dolphin, who appeared no worse for the experience. In fact, Primo did a couple of victory laps around our boat as if to say, "Now I'll show you how fast I can really swim."

On the way back to the dolphin pens, as Primo leaped and played in the boat's wake, I untangled the mess that was once my $3,000 dolphin heart rate monitor. It didn't take long to piece together what had happened. With remarkable aquatic precision, Primo had used the outboard motor propeller to cut off the harness and the monitor. While the dolphin escaped without a mark, he managed to slice through the nylon webbing and cut the suction cup wires. In addition, the metal casing of the microprocessor had three propeller strikes on it that destroyed the instrument.

I couldn't be mad—it was all too incredible. Instead, I marveled at the intelligence of the unencumbered dolphin swimming with ease next to me.

When I hooked the heart rate microprocessor to a computer, I found that the data was intact. It revealed how dolphins lowered their heart rates when bow wave riding. I wrote a paper on our serendipitous discovery, which was turned into a lead article in the international scientific journal *Nature*. Primo in all his leaping glory was the cover boy on the issue headlined "Why Dolphins Hitch a Ride." Primo didn't ruin my science. Instead, he had demonstrated that my hypothesis was wrong.

I'VE KEPT the battered microprocessor monitor casing on my desk as a reminder of instinct versus intelligence in animals, and I took Primo's scientific lesson to heart. Since that incident, as well as those with Austin and many other creatures, I've relied on animals to lead my science instead of the reverse. It would be the same with KP2, and it was guaranteed to be much more interesting than I ever could have imagined.

14.

Breath by Breath

There were so many pressing questions that needed to be answered for Hawaiian monk seals. Were there enough fish in the oceans around the Hawaiian Islands to satisfy both the monk seals' and human appetites? Was climate change impacting the ability of Hawaiian monk seals to survive? Why were so many young seals dying? And most important, would we learn the answers fast enough to keep this incredibly unique tropical species from disappearing forever?

The first experiment I had in mind was not a fancy, high-tech molecular showstopper. Instead it was a time-tested 1940s technique with a few variations to accommodate KP2's aquatic habits. We were going to measure the seal's metabolism as he rested in water. Just as our own metabolism determines how much each of us needs to eat and dictates whether ingested calories go to heat or onto our hips, KP2's resting metabolic rate would indicate the minimum number of fish it takes to sustain a monk seal. With that one number I could begin to match the monk seal's metabolic needs to the fish available in the Hawaiian Islands. I knew this was literally entering into very dangerous waters, for the answer could very well pit the livelihood of Hawaiian fishermen against the survival of Hawaii's monk seals.

There was no predicting what we would find. By living in warm waters, the Hawaiian monk seal was different from any other phocid seal. They were the only pinniped living in the islands, and I wondered if the tropical seal species had set its thermostat on low in comparison to temperate-living or polar-living seals, which needed to digest lots of fish to keep warm.

For KP2's FIRST EXPERIMENTS, we created a spa pool and metabolic chamber next to the big sealarium pool. We scavenged an old plastic tub and mounted a skylight dome on top to create the airtight chamber. By adjusting warm salt water and ice we were able to simulate Hawaiian temperatures in the spa, from the coldest that occurs below the oceanic thermocline, where monk seals dive for fish, to the warmest waters that make shallow tropical lagoons feel like a Jacuzzi. To ensure that we didn't train KP2 to ramp up his metabolism in anticipation of a particular temperature, Traci randomized the spa water temperatures; every test day would be a surprise for the seal and for me.

All we needed was KP2's cooperation. As much as the young seal seemed to thrive on the human attention, he had other ideas when it came to joining us at the spa pool and sitting still. There were many days in the beginning of his science training when I could not tell if

I was asking too much of the seal, if he was playing the obstinate two-year-old, or if he was just not that bright.

For weeks Traci and Beau tried to get KP2 to understand the research task before him. I'd watch from the sidelines as over and over again the young seal played in the spa like it was his private bathtub. We could barely see the seal for all his splashing.

"Are you sure he can do this?" I repetitively queried the trainers every few days, convinced that we had finally met our match in KP2. Puka and Primo, as well as Wick, Morgan, and Taylor, had all figured out the metabolic game. The dolphins and the sea otters quickly learned to float quietly on the water, enthusiastically reaping all the fish rewards for being asked to do so little. KP2, with his boundless energy, scuttled every attempt at science.

His first day in the spa he dove with his tail stuck up in the air and hind flippers whirling. He explored every inch of the small pool, circling and splashing until he'd created a whirlpool. He sniffed and dove, constantly looking for something to butt his head against or chew on. When tired of exploring the bottom of the pool, he came to the surface to roll upside down with his belly button exposed. Then with a snort he performed a somersault. KP2's fun time repertoire seemed endless.

"Hoa, give it a rest . . . please?" After several weeks Beau was on his knees in front of the spa practically begging the whirling seal to calm down. This was not a position any animal trainer wanted to be in. Once you resorted to pleading, you'd lost the game. KP2 had effectively demoted Beau to one of those dog owners being dragged down the street on the end of a leash imploring, "Stop, Cookie, stop!" Unfortunately, I had placed Beau in this terrible predicament. He was stuck between KP2 with all his two-year-old energy and me with my demanding research schedule.

Beau sighed but appeased me by lowering the dome on top of the spa and splashing seal. For a brief moment the seal stopped moving.

"There—see? All he needed was a little distraction," I gloated, and began taking oxygen measurements from the dome air outlet. With fresh air pumping in through a hose, KP2 became fascinated with the stream of air on his whiskers and face at the inlet. All too soon, the seal had his muzzle jammed into the end of the hose. The vacuum pump supplying the air at two hundred liters per minute screeched. I could hear the motors strain and smelled the burning of bushings. With KP2's muzzle wedged in the hose, the pump began to create a vacuum in the dome and the water rose quickly.

"Stop! He's going to flood the pump!" I yelled. In an instant, Traci had the skylight dome raised, and the water receded. KP2 watched all the frantic activity around him with the same nonchalant expression he had in Hawaii when he was the center of chaos at the Kaunakakai Wharf. As the commotion finally settled, the seal looked up at us and rumbled a robust "*brrrrauughhrrr!*" He informed my team that it was time for his breakfast.

We were getting nowhere. But I had seen it all before, and from a much more formidable marine mammal than the young seal. The most creative and one of the most intelligent animals that I'd placed under a metabolic dome was Shouka, the female killer whale from Six Flags Discovery Kingdom in Vallejo, California. At six thousand pounds, she was the biggest animal I'd ever tried to measure.

For Shouka, the skylight dome was fittingly huge—over eight feet long on either side and four feet in height. It took two people to maneuver the dome into the killer whale's pool. We used several Sears shop vacs plumbed in series to pull over five hundred liters of air per minute through the giant dome to accommodate the size of Shouka's explosive whale exhalations.

Almost immediately Shouka had begun inventing games during our metabolic experiments. First she developed a habit of sticking her massive pink tongue out to lick a plastic float meant as a chin rest. Then Shouka chomped down, crushing the white float, ripping it from the post, and tossing it to the feet of her startled trainers. In the following sessions the killer whale began to spit water from her blowhole, coating the inside of the dome with orca gusto. When not spitting, she began humming. The sound grew louder and then shifted into the high-pitched *"eeeeeeee"* groan of a creaking door opening slowly. There were squeaks and rumbles, creaks and pops, whistles and hums.

Through it all, we kept monitoring the whale's metabolism and discovered that there was a measurable energetic cost to orca vocalizations. That finding had immense implications for wild killer whales that vocalized when disturbed by ship traffic. Several years later, scientists from the National Marine Fisheries Service used this information to conduct a larger study regarding the effects of boat traffic noise on killer whales and their impact on the salmon populations of Puget Sound. Without ever refusing a session, Shouka had changed the course of killer whale science.

It was this type of surprise that made animals so fascinating to study. As frustrating as it was for Beau and Traci, I knew that if we were patient with KP2, he would eventually show us something amazing. It didn't take him long to prove me right.

ONE DAY, SIX WEEKS LATER, Beau and Traci called me down to the spa pool.

"Water," Beau called cheerily to KP2. The seal immediately hauled out of his sealarium pool, inchwormed his way down a connecting walkway, and slipped in his typical molasses ooze over the edge of the spa pool. He took a quick look around underwater and then popped his head up, waiting for the next challenge.

"Chin," Beau instructed KP2. The seal gently set his chin on a bar in front of him and closed his eyes, resting peaceably.

The first time I saw this I was astounded.

"How the heck did you get him to do that?"

"I just figured he needed a little help," Beau replied matter-of-factly while watching KP2 float quietly in the spa. "He has so much blubber that he was having trouble floating in one place. Since he already knew how to rest his muzzle on a stand for his eye exams, I just added a new chin rest to his spa."

Beau was being too modest in his assessment. In truth he and KP2 had come to an agreement. The breakthrough had occurred during an especially long training session several weeks back. It had been an exhausting session for both the seal and the trainer, one in which, for the hundredth time, Beau asked the seal to enter the spa pool and float for ten minutes without diving, without swimming or splashing, without vocalizing or trying to climb out. On this one day Beau perceived a flicker of understanding in KP2. Whether the seal was trying to disarm Beau or finally understood the chain of behaviors can be debated. Regardless, the light finally went on in his brain.

To KP2, the chin rest was like floating on a new kind of boogie board, a habit that had stayed with him from his first days in the rehabilitation pools at the Kewalo Research Facility and his frolicking good times with paddlers at Kaunakakai Wharf. He eagerly rested on it. Once the seal was in position, Traci added one last piece of equipment: the skylight dome.

Under the clear plastic, KP2 watched Beau intently, never letting his eyes wander while resting in the water. By continuously taking samples of the exhaust air from the dome, I began to measure KP2's metabolism using a calibrated oxygen analyzer, the one piece of high-tech equipment involved in the tests.

"People pay big bucks for this in fitness centers," I informed the seal. KP2 yawned back. The trainers and I knew that the test would last only as long as the seal felt like cooperating. It was his choice to stay on the chin rest or to leave.

For some animals, leaving the skylight dome could be quite spectacular. Big males never seemed to understand the strength of their bony skulls. Woody, a thousand-pound adult male Steller sea lion, and Sivuqaq, a seven-hundred-pound male Pacific walrus, both managed to accidentally pulverize skylight domes with their enormous blocky heads when they inadvertently backed up. In the Antarctic, a male Weddell seal once mistook the skylight Plexiglas for a thin layer of ice. After a couple of trial pushes using his nose to press gently on the skylight, he submerged and then explosively popped up, breaking a clean hole in the dome with his muzzle. Marine mammals, especially male marine mammals, liked to use their thick skulls as battering rams, which wreaked havoc on my metabolic equipment.

To prevent KP2 from doing likewise, Beau had discovered a solution to the seal's boredom. Borrowing from Wiki, the cockatoo, the trainer began to bob and dance in front of the spa pool. Whenever KP2's eyes began to drift, the trainer in Hawaiian board shorts broke into a dance. He quickstepped and shuffled. He jumped and swayed. The trainer did everything to keep the seal's attention for the entire metabolic session. After fifteen minutes Beau clinched a perfect session with a country-western two-step.

"Done!" I shouted from behind the oxygen analyzer. "We got the data point!" Traci immediately raised the skylight dome while Beau began feeding KP2 fish as fast as the seal would swallow. We finally had our first metabolic data point for a Hawaiian monk seal—a data point sixteen years in the making, if you counted the time it took to obtain a permit.

"It's 5.07 milliliters of oxygen per kilogram of body mass per minute," I announced proudly to the team. "That translates to twenty-eight hundred calories a day for a resting seal—about the same as a moderately active human." The numbers were not as important to the others as was the completion of a successful session. For the first time everything, from the equipment to the people to the seal, had gone perfectly. It was a historic day for all, and with that Beau finally collapsed in exhaustion.

Back in my office I recorded our first data point in my notebook and created a graph, comprising a lone circle on a white field, that showed KP2's metabolic rate in relation to water temperature. My mind spun with questions, now that I had a starting point. KP2's metabolic data point was high in comparison with adult Weddell seals but low compared with Weddell pups. Could that mean that tropical seals had turned down their metabolic thermostat? Does that make monk seals better or poorer divers? How would water temperature affect the seal's metabolism?

"Now all we have to do is fill in the rest of the points," I said to myself. Energized by the prospect of embarking on a new scientific scavenger hunt, I decided it was time for a celebratory jog down the beach.

MEANWHILE, BEAU WENT BACK to the sealarium with KP2, both of them weary from the day's activities and concentration. Under the humid canopy he started talking to the seal.

"Thank you for your hard work," Beau said quietly.

KP2 sat in the water listening to the trainer who had spent so many hours working with him day after day.

"I hope your journey through this project will be enjoyable," Beau continued. He spoke of everything and nothing to the attentive seal. In a one-way conversation he discussed their individual journeys, their past, present, and future, and ended on what he planned to do for the rest of the night.

Beau had no misperceptions about what the monk seal actually understood in terms of his words. Rather, the trainer recognized the one thing that KP2 loved: people. KP2, as a hand-raised pup, thrived on human company. The sound of their voices was as soothing as any tropical wave washing over his body. So the trainer gave himself and all his inner thoughts to the seal that day as a special reward for his patience under the skylight dome.

When Beau finally ran out of words, KP2 hauled out onto the deck next to him and fell fast asleep. Maybe his choice of sleeping spot just happened to be where the sun was streaming in, or maybe it was the comfort of Beau's presence, but a deep connection was forged between man and seal in that quiet moment.

With a partnership forged in friendship and fish, my team began to unravel the secret biology of monk seals, a biology that had remained hidden for over fifteen million years. I found it both remarkable and depressing that we had come to such a desperate crossroads for this species. We were in a race to discover everything we could about these Hawaiian pinnipeds before it was too late for the species. Every data point for KP2 represented another piece of the monk seal puzzle and another clue to his species's survival.

Once KP2 mastered the art of sitting still, we began to add environmental variations. In one of the first series of tests, we determined the effect of water temperature on his metabolism. How he responded would help delineate where along the Hawaiian archipelago monk seals should raise their young. The poor survivorship of Hawaiian monk juveniles made locating the ideal environment for their early years a priority. In the face of rapid changes in ocean temperatures and ever increasing pressure from human activities, carving out a private place for monk seals was becoming increasingly difficult. Knowing the preferred environments of the seals was the first step in developing biologically meaningful sanctuaries.

Early each morning, Traci filled the spa pool with the experimental temperature of the day. To keep us honest, neither KP2 nor I knew whether she was going for a chilly deepwater thermocline or tropical lagoon warmth. It didn't take long to guess the pool temperature. KP2 let us know loud and clear what he thought of Traci's daily choice, even without the benefit of my expensive oxygen analyzer.

"*Snnneeww*," KP2 exhaled in long, slow breaths as he floated luxuriously when the spa pool was filled with 86°F water. The seal was completely relaxed when Traci cranked up the temperatures. Soaking in the warmth, he was back in the islands relishing the heat from his private cove; the warmer the water temperature the better, according to the behavior of the small seal.

The reverse occurred on the mornings that KP2 dipped into the pool set at diving temperatures. He'd dunk his nose to test the water, look back up at Beau, then the water, and then Beau again.

"Water, Hoa," Beau called in his usual encouraging voice. KP2 was no fool and was unmotivated by the trainer's false cheeriness.

Despite his obvious reservation, KP2 always slipped into the water even if it was chilled to 60°F. His body language and the speed at which he entered the spa told me the water temperature more accurately than any calibrated thermometer. On the chilliest mornings KP2 would slowly ooze into the cold water and then add a Three Stooges "*whup, whup, whup*" to voice his displeasure before the metabolic measurement.

KP2's cooperation paid off when we compared his metabolic rates for all of the water temperatures tested. When the data points were plotted side by side, it was apparent that temperatures below 60°F challenged the young seal and instigated a rise in his metabolic rate. This was our first clue that deep diving came at a high cost to young Hawaiian monk seals. Habitats where fish resided in cold, deep waters would be a caloric tightwire act for these animals as they tried to balance the metabolic calories expended to keep warm against the calories ingested when they finally caught a fish.

With overfishing and changes in fish distributions as oceanic temperatures fluctuated, the caloric challenge had became dire for monk seals as they swam farther and dove deeper to find food. The young seal's warm-water metabolic biology also revealed another important clue about their survival. Based on our findings, it appeared that Hawaiian monk seals were locked biologically into tropical paradise. Unlike highly mobile marine mammals, such as humpback whales, that visited the islands each winter to breed and give birth but then migrated thirty-five hundred miles to the cold, nutrient-rich waters of Alaska to feed, Hawaiian monk seals were oceanic prisoners due to a biology that demanded warmth. As such the seals had to make peace with the islands' native fishermen or go extinct.

There was more to the unique biology of the Hawaiian monk seal than simple metabolism. Internally, monk seals were built differently than any of the other seals or whales I'd studied. During necropsies to determine the cause of death in stranded marine mammals, I had always been amazed by the extraordinary length of their small

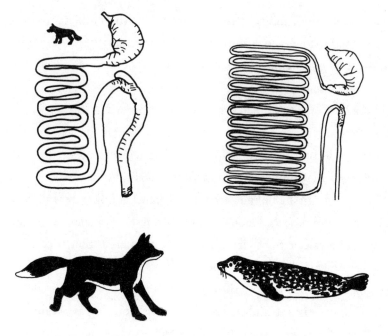

intestines. The guts of dogs, humans, lions, and other terrestrial mammals looked like coiled bicycle inner tubes. In contrast, the intestines of marine mammals appeared as yards and yards of balled yarn stuffed into their abdomens. Inspecting the guts during a necropsy, I looked like a magician pulling a never-ending trail of scarves out of a top hat.

The guts were so impressive that I began to measure small-intestine lengths in as many different species as possible, and found that exceptionally long intestines were a defining characteristic in marine mammals. Retriever dogs have a 9-foot small intestine while a similarly-sized harbor seal has an amazing 49-foot intestine. The human small intestine is 16.5 feet long while a dolphin's is 102 feet and the killer whale's is a whopping 177 feet long. End to end an orca intestine spans the height of the Statue of Liberty from base to torch and then some.

I hypothesized that such extraordinarily long guts were important for heating marine mammals. Water robs a mammal's body heat twenty-four times faster than air at the same temperature. That is why submerged humans quickly succumb to hypothermia even in relatively temperate waters. Marine mammals solved this problem with a thick insulating blubber layer coupled to an internal heater. The act of processing food provided the heat; long intestines allowed that process to occur fast enough and long enough to keep the animals warm even in the chilliest of Antarctic waters. But what if the marine mammal lived in hot water?

The answer was in the guts of monk seals. The small intestine of a monk seal the size of KP2 is thirty-six feet long, four times that of a dog but thirteen feet shorter than the intestinal length of its cold-water-dwelling harbor seal cousin. Among marine mammals, it seemed that a warm-water lifestyle had permanently restructured the internal organs of the monk seal, thereby creating another barrier for this species ever to leave the islands. More than any other marine mammal species in the islands, the Hawaiian monk seal was biologically linked to the warm waters of paradise.

By my calculations, they could never survive the long, cold-water swim that would be required to reach another home if humans refused to share the ocean.

15.

Killer Appetites

N o sooner had we collected and published our data on KP2's metabolism than it was used against him and his species. From the very beginning not every islander embraced KP2 or the ever increasing presence of his species in the main Hawaiian Islands. Rather, the encroaching monk seals with all their caloric needs were viewed as competitors for dwindling oceanic resources. To make matters worse, things were changing rapidly in Hawaii's coastal ecosystems, just as they were all over the world. Local lobster fisheries had crashed in the late 1980s and the groundfish fisheries were in trouble. Recreational and commercial fishermen saw in monk seals the final nail in the coffin of their faltering business. Making a living on the water for sport or sustenance had become increasingly difficult in Hawaii. With the double knot of rising fuel prices and collapsing fish stocks, local fishermen felt an ever tightening noose choking the life out of their livelihood, a practice that had been a part of their island families for generations.

I empathized with those who lived on the water. My life had been spent in the company of fishermen. My father had fished the waters of Chesapeake Bay. My brother still fished the bass lakes and coasts of South Carolina and Georgia. With the exception of my mother, who could never bring herself to move farther than ankle deep along a beach, we all had succumbed to the lure of the ocean. Salt water truly flowed in our veins. However, we loved the seas from opposite ends of a fishing pole, which could create friction on a boat.

"How can you be a fisherman and believe you are not affecting the ecosystem?" I'd ask.

"How can you be raised a Catholic and believe in evolution?" my father would retaliate.

Some things defied explanation.

I did not possess the soul of a fisherman. They had a rabid passion for catching fish that seemed spurred on by hidden demons, as if someone else might hook a fish before them. I didn't understand their devils nor could they understand the ease with which I'd allow a sea lion, dolphin, or shark to steal a fish off my line.

We'd spar as families do, with no resolution. In their minds the ocean's bounties were endless and theirs to exploit. In my mind we had reached a point where sharing was essential if we all expected to survive.

CLEARLY, IF KP2 AND HIS species were to survive, they would have to obey nature's biological rules as well as man's territorial rules. In modern Hawaii this had developed into a tenuous situation as seals and humans alike overstepped the ancient line separating them. KP2 had entered into the human realm where no wild animal belonged; he compounded the transgression by befriending the swimmers and surfers of Kaunakakai Wharf. His rescuers, including myself, had crossed into the seal's world and violated the cardinal rule of allowing nature to take its course; we had intervened when a seal mother and nature dictated that KP2 should have died the day he was orphaned.

In recent years the population of monk seals had slowly expanded southward, leaving the remote Northwestern Hawaiian Islands to arrive on the beaches of Kauai, Molokai, Oahu, Niihau, Lanai, Maui, and the Big Island of Hawaii. Fourteen years earlier when I'd lived in the islands, a monk seal sighting was a cause célèbre, a rare occasion filled with wonder and excitement as these exotic creatures hauled their quicksilver bodies out of the surf. Since then the number of monk seal mothers giving birth in the main Hawaiian Islands had increased annually. The animals arrived unannounced, gave birth, and then disappeared abruptly six weeks later once their pups were weaned. Locals were afforded only a brief glimpse into the monk seals' world. The problem now was that more and more of the tourist seals, like their human counterparts, were deciding to stay.

Based on our metabolic findings, there was no denying that young monk seals had a healthy appetite for fish. Compared to adults, juvenile monk seals were like growing teenagers. A growing monk seal's caloric demands translated into a choice of twenty-three pounds of octopus, ten pounds of eel, or sixteen pounds of spiny lobster per day to survive in the wild. For some island fishermen, especially those who had spent a lifetime on the water, this level of gluttony was intolerable. Consequently, while KP2 was making scientific discoveries in my lab on the mainland, his wild cohort was being attacked.

In December a twenty-year-old male monk seal, renowned for

his enormous size and missing hind flipper due to an earlier shark bite, was found floating in the waters of Kaunakakai where KP2 once played with the local children. Leimana Naki, a paddler, made the grisly discovery and towed the seal in by kayak. A necropsy performed by government officials determined that the seal had been intentionally killed but did not elaborate as to how the animal died.

The Kaunakakai seal was the latest incident in a disturbing trend. Seven months earlier on KP2's birth island of Kauai, a seventy-eight-year-old local had hoisted a Browning .22 caliber rifle to his shoulder and fired four rounds at a tagged female monk seal known as RK06. Two of the rounds found their mark, killing the fifteen-year-old pregnant female in the shallow waters near Miloli'i Beach. Her intention in arriving at the Garden Island had been to give birth to her male pup. The elderly man's intention had been to scare her out of the area to prevent competition for fish. He had failed to recognize that female seals rarely ate during the nursing period; their pups fed only on milk.

In the middle of this conflict between monk seals and local fishermen, federal scientists proposed a radical idea. In the Hawaiian Islands nearly all the monk seals born the same year as KP2 were already dead. Oddly, some of the worst-hit areas were not where one would have predicted. Rather than the heavily trafficked waters of the main Hawaiian Islands, it was the most pristine Northwestern Hawaiian Islands, situated in a newly designated marine reserve, that represented the death trap where pups and juveniles were disappearing in vast numbers. To correct the problem, Dr. Charles Littnan, the lead scientist for the Hawaiian Monk Seal Research Program under the NOAA Pacific Islands Fisheries Science Center, and a team of federal and nonfederal scientists proposed moving young seals into the fishermen's oceanic backyard. As outlandish as it sounded, their logic was simple.

The Papahānaumokuākea Marine National Monument, established in 2006 under the George W. Bush presidency, safeguarded over 105,564 square nautical miles of tropical water surrounding the Northwestern Hawaiian Islands. As such it created one of the largest conservation areas in the world. All commercial and recreational fisheries were banned, leaving the beaches, waters, plants, and animals to the whims of nature.

Yet government scientists found that nature was frustratingly capricious. Inexplicably, the Hawaiian monk seal failed to thrive in Papahānaumokuākea. Ten years ago pups born in the remote tropical atoll of French Frigate Shoals within the reserve were almost guaranteed to see their second birthday. Today a mere 8 percent of the pups survived past this age.

Through a series of eliminations and scientific observations, Dr. Littnan's team had identified two likely culprits: entanglement in marine debris and competition from larger, more aggressive fish. Whereas entanglement had been a long-term problem for the curious seals, competition for food was relatively new. Like diminutive cheetahs being chased off their kills by burly hyenas, juvenile monk seals were being bullied off their fishy meals.

A National Geographic Crittercam worn by several of the wild seals had revealed the problem. Left to their own devices, Hawaiian monk seals cruised the shallow blue waters and reefs of the outer islands. The remarkable videos showed that the seals used their blocky heads like battering rams. They foraged by sticking their muzzles into reef crevices and by head butting rocks and pieces of broken coral just like KP2 butted coconuts and his bowling ball. Small fish hiding in the coral were scared out into the open, where they could be easily snatched up by the hungry seals.

But the wary eyes of other carnivores were watching the monk seals' fishing activities. Large numbers of predatory sharks and ulua fish, also called jacks, followed the monk seals on their dives. They hung in the shadows until the seals scared up a fish. Before the seals had a chance to catch their prey, the larger predatory fishes sprinted in to nab the fleeing meal. Weaned pups and juvenile seals were especially vulnerable to this parasitism. Too slow to compete, they were often left to go hungry after a long day of hunting. The emaciated bodies of young seals now littered the shores of the Northwestern Hawaiian Islands.

In an unorthodox, seemingly desperate management plan, the government scientists proposed to solve the problem by translocating a cohort of newly weaned female monk seal pups. Immature females would be taken from their birth sites in the remote Northwestern Islands, where they had no defense against marauding sharks and jacks, and then shipped to the "big city" of the main Hawaiian Islands to mature into adults.

Ironically, the presence of commercial and recreational fishermen in the main islands had done the seals a favor by targeting the large predatory fish that competed for the seals' prey. Bigger fish brought in more money. As a result, the large, aggressive fish that plagued pups and juvenile seals in the outer islands were gone. In the main Hawaiian Islands, young seals were free to eat and grow. If the management plan was successful, the teenage female seals would be returned to their original birth sites in the Northwestern Islands several years later, once they were large and strong enough to compete with the predatory sharks and jacks.

However, the bold plan required one critical element for it to work: people. The locals in the main Hawaiian Islands had to be willing to share their waters and beaches with the monk seals. The loudest voices in the ensuing town hall meetings to discuss the plan indicated that the prospects were not good.

On Maui the locals objected to the government and mainland scientists telling islanders what to do, likening Hawaiians to the true endangered species in the proposed scenario. Monk seals were deemed
raiders of their "ocean icebox." Oahu fishermen echoed the sentiments
and criticized the move, citing the loss of fishing gear to marauding
monk seals. "The government should not be trying to play Mother
Nature," they told the NMFS representatives. At the town meeting on
Molokai one resident even brought a cooler full of fish to demonstrate
how much a seal would eat in a day. "We don't even eat that much in
one month!" the woman declared. To my horror, I realized that the
metabolic rates and caloric demands that we had just determined for
KP2 were being used against his species.

Locals pitted seal appetites against individual human appetites.
But the woman's cooler demonstration was not quite accurate. It was
true that a growing monk seal ate a lot of fish—eight to nine pounds
of fish per day, in the case of KP2. However, that fish represented *everything* the seal would eat and drink in twenty-four hours, since both
calories and water came from the seal's diet.

We had discovered that the monk seal's caloric demand was between that of a teenaged Hawaiian surfer and an Olympic-class swimmer such as Michael Phelps, depending on how active the seal was. In
human terms, the surfer would need to eat more than ten In-N-Out
burgers (at 480 calories per burger) per day to remain in caloric balance if that were the only thing he ate. Michael Phelps would need to
consume twenty-five burgers during peak training periods. An active
juvenile monk seal would need seventeen and a half burgers (although
I admit we did not test this on KP2 when we stopped at In-N-Out on
our drive through Los Angeles when he was a pup). In terms of calories, pound for pound teenage seals and human males who played in
the water were not all that different. Adding a cohort of female juve-

nile monk seals to the main Hawaiian Islands, as Dr. Littnan's team proposed, could be likened to an influx of young surfing tourists.

By comparison, larger marine mammals were high-end consumers. It would take 46 In-N-Out burgers to satisfy a 440-pound dolphin's daily appetite and an astounding 354 burgers to feed a 6,500-pound killer whale each day. Unlike the oceanic cetaceans, the monk seals were making the mistake of eating in front of the fishermen. Dolphins and whales, on the other hand, seemed to forage in comparative peace offshore.

I was curious. Were monk seals as gluttonous and fat as some fishermen accused? We had been weighing KP2 nearly every week since he'd arrived. Yet that was not enough to answer the question. We needed to measure the seal's blubber thickness. To do so, we used a portable ultrasound unit, the same instrument found in ob-gyn offices to provide eager parents with a first glimpse of their son or daughter in utero. For us, ultrasound proved to be a wonderful, noninvasive means of seeing and measuring the layer of insulating fat that encased seal bodies. It was the newest technology for assessing one of the oldest and most important metrics of animal body condition: how much fat was on board.

Before ultrasound technology, the methods for evaluating fat in wild animals ranged from distasteful to deadly. One option was to sedate the animal and then use a biopsy punch on the immobilized body to scoop a wedge of fat from beneath the skin. For a big whale, this could mean nearly twelve inches of coring. The alternative, old-fashioned naturalist method was simpler but did not do much for the health of the animal. They shot first, then skinned the carcass to reveal underlying fat deposits.

Squinting at the blurry ultrasound images on the tiny black-and-white screen, I tweaked the gain knob to focus on KP2's fat layer. He was being especially patient as I fiddled with the instrument. Under Beau's and Traci's watchful eyes he was content to let me run the ultrasound probe across various body sites, except for his belly.

"Roll him, please," I asked Beau.

With that, Beau gave KP2 a signal to roll over onto his back so I could measure his belly fat. For most marine mammal trainers this meant showing a hand and turning it from palm down to palm up. As soon as Puka and Primo saw the trainer's hand turn, they immediately turned upside down with their pink bellies pointed to the sky. To accommodate KP2's poor eyesight, Beau added a voice command just to make sure he got the signal: "Hoa, over!"

Having warmed to his sunbathing spot, the seal initially pretended not to hear the signal. Then, in tropical slow motion, he rolled leisurely onto his back. When I slathered blue ultrasound gel onto his white belly, his whole body contorted. When I tried to slide the ultrasound probe near his belly button, KP2 curled into a "C." And so I made two discoveries: A healthy young Hawaiian monk seal has a one-inch-thick blubber layer on its belly. And, at least in the case of KP2, they are ticklish.

I found the monk seal's thick blubber layer a mystery. Why would a seal species designed for living in the tropics need so much insulation? Admittedly, their blubber was only half as thick as that of a Weddell seal's, but it was 100 to 150 degrees colder in the Antarctic seal's icy home. The reason for KP2's substantial blubber layer was likely related to padding for body streamlining or as a fatty food store to sustain the animal in lean times. But we would have to await another experiment at another time for proof.

While we were busy studying KP2's blubber and trying to determine the relationship between the caloric needs of seals and fishermen, an article arrived from the *Molokai Dispatch*. Walter Ritte had recognized that the town hall meetings in the islands were not going well for the monk seals, and he decided to do something about it. The former protester at the Waikiki Aquarium enlisted students from the Hoʻomana Hou School to create a video about preserving Hawaii's seals. In an odd twist given the current competitors of wild seals, the fictional account starred two sharks and a group of fishermen who came to the rescue of KP2.

"What are they rescuing KP2 from?" Traci asked.

I paused as I read the article.

"Well, according to the newspaper . . . ," I replied slowly, "I think it is us."

16.

The Social Seal

W hat are we doing here?" I asked Beau and Traci in frustration when I read the press release about KP2's rescue video. I was still reeling from reports of the latest monk seal killing in Hawaii, and from my own classroom discovery. As a part of my Comparative Physiology course I had brought forty unsuspecting undergraduate students into the campus necropsy lab, where we performed a CSI-type examination of a sea lion carcass that had washed up on our beach. Usually, these cases were a simple shark bite, parasite, or lung infection. Today was different.

"Your goal today," I informed the students, "is to determine how this supposedly healthy California sea lion died."

Lying in front of them was a beautiful full-grown adult male sea lion that appeared to be in the prime of his life. Dark, sleek, and massively blubbered, he was an impressive animal. It saddened me to see him unceremoniously draped on a stainless steel table. He was quite dead without a mark on him. Even the attending veterinarian was stumped as to the cause of death.

The first midline incision sent a number of the weaker stomachs out the door, as blood poured from the abdomen of the sea lion. Something was very wrong. As soon as the animal's internal organs were exposed, the problem was immediately apparent: lying in a small indentation on the liver was a bullet. He had been shot with a small-caliber handgun. The bullet had penetrated the sea lion's blubber without leaving a mark, severed a splanchnic artery, and come to rest on the liver.

I was shocked and then visibly angry, emotions I never reveal in a classroom.

"What a waste," I told the students disgustedly. It was salmon season off the coast of Monterey Bay. With the fishery shut in previous years because of low fish stocks, the competition for the pink meat was heightened this year. Recreational and commercial boats were out in force on the waters, and sly California sea lions had learned to follow them to nab the fatty salmon off their fishing lines. This year there was no mercy shown to the ravenous marine mammals. At times the fishing grounds five miles offshore sounded like a war zone, with the incessant *pop, pop, pop* of faraway gunfire. Whether to scare the animals off or to take them out permanently, this bullet had found its mark. I wondered how long it would be before the California sea lion was in the same sad state as the Hawaiian monk seal.

"WHAT IS THE POINT of our science if people are going to hate us and shoot monk seals into extinction?" I asked my trainers.

"Well, I thought KP2 was supposed to be a sign from the ocean," Traci reminded me. "Why doesn't *he* say something?"

I had never been in a situation like this. My science was being drawn further and further into social and cultural circles, and I was unsure how to react. In Antarctica there were no local residents, fishermen, or politicians to consider. Data were just data with no ulterior motives or families at stake. In Hawaii, nature and an ancient culture were intertwined. Especially in the islands it was critical that locals got the science right, that they understood where their families and the families of wild monk seals fit into the entire tropical ecosystem. We had to do more than simply generate scientific data and go on our way. We needed to communicate with the people, a prospect I found exceedingly frightening as a reclusive scientist.

THE ARTICLE in the *Molokai Dispatch* included a photo showing a large group from town gathered at Kaunakakai Wharf, KP2's favorite old swimming spot. Locals had mistakenly heard that the real KP2 was arriving back on the island, saved by fishermen from a life in California. Instead they greeted a youngster in a gray seal costume meant to look like KP2. The "seal" waved a flipper next to two papier-mâché sharks in a boat that had just traveled from my lab.

Being upstaged by paper sharks and fishermen would have seemed funny had it not been for the rescue by the scientists in California. I felt the need to set the record straight, if for no other reason than to defend my fellow researchers. People had to understand why KP2 was here.

That night I wrote a letter to the editor of the *Molokai Dispatch*. Trying to dispel the myth concerning scientists, I wrote, "As opposed to white lab coats my research team goes to work in flowered board shorts and flip-flops." I told of KP2's daily life and work with us. Most important, I acknowledged that the elders of Molokai were wise when KP2 was blessed with the name Ho'ailona. "This seal does carry a message—that all of us—islanders and mainlanders, school-children and adults, fishermen and scientists alike—must live together and share the oceans."

The response from Molokai was a series of heartfelt e-mails and phone calls from KP2's former caretakers, the palm frond shakers, and the general public. So many people were excited and curious about KP2's foray into science that I was encouraged to say more.

I decided it was time to follow in the long tradition of Hawaiian communication through storytelling and gave the seal his own voice. It had to be an exceptionally big voice in order for it to carry across the Pacific Ocean, so I had KP2 join the newest form of social communication: Facebook.

With a doctored age so that he was old enough to join Facebook, KP2 began talking to the children of Molokai, the aunties and uncles who'd once cared for him, as well as anyone willing to friend him. To cover all bases, his name was Ho'ailona Monk Seal (Kptwo Monk Seal). No subject was too small or too big for him to discuss. He told his friends about his early abandonment, about his love of the oceans and Hawaii, about environmental issues and his life in Santa Cruz. He spoke of the oceans in peril, of overfishing and pollution, of climate change and the impact of oceanic sounds—and he was not above commenting on such important topics as the joys of In-N-Out burgers or the Giants' World Series win (although I made sure he was also a Yankees fan like me). He was a single male who worked at the University of California, and his residence was listed as "Sealarium."

Almost immediately I ran afoul of island sensitivities as the Facebook voice of KP2 reverberated through Hawaii. Searching for a way to describe the shape of the seal, I innocently had him ask his four hundred new friends, "Some of the folks here say I look like a banana slug. What do you think?" I signed it "Your Hoa." The response from the islands was swift and brutal. Through e-mails and KP2's Facebook page I was harshly scolded for my "insensitivity," "disrespectfulness," and "immaturity." Complaints sailed from the islands, across the mainland, and to the NOAA and National Marine Fisheries Service permit offices in metro Washington, D.C. By likening the seal to the UC Santa Cruz school mascot—the Banana Slugs—I had created a transoceanic national incident.

"It was meant to be funny, an honor. John Travolta wore a UC slug T-shirt in *Pulp Fiction*. It's our school mascot!" I defended myself to David Schofield when he called from Honolulu to explain the trouble I had caused. David had been in my position many times when working with stranded seals and whales in Hawaii.

"You have to understand. These ocean animals are *'aumakua*, a physical manifestation of nature, of deity in Hawaii," he responded seriously.

I felt stupid. Had I just likened a god to a worm? No wonder people were upset. For traditional Hawaiians there was no distinction between nature and culture; the *'āina* (land), the *kai* (ocean), and the *lewa* (sky) were intertwined in the life and spirit of the islands.

Still, I argued, if monk seals were *'aumakua*, how could anyone be allowed to kick or shoot them? I didn't understand. Obviously, my ability to read people fell considerably short of my skill with animals. I immediately removed the offensive posting and from then on chose KP2's words with the utmost care.

. . .

SOON OUR LIFE WITH KP2 settled into a pattern as we entered into the warmer days of spring. The trainers opened the doors connecting KP2's sealarium to the outside decking, and he wandered around according to the shifting sun to choose the warmest, most windless spots for his daily suntan sessions. The cockatoos noted his appearance with a brief display of squawking and then went on with their day, talking to themselves.

In between the seal's naps Beau and Traci continued his training for our research. We attempted to learn everything we could about the monk seal, from seasonal changes in his blubber layer to growth patterns to his metabolic rate as he swam loops around the perimeter of his pool. Through it all he was a keen participant, always greeting each day and experimental session as if it were a game. Life was good—that is, until April blew in.

"KP2's shut down."

"Shut down how? Shut down as in he is sick?"

"Well, not exactly."

"Shut down as in he is too fat to care?"

"He is . . ."

"Shut down as in he just doesn't want to do anything you are asking him to do?"

I went down the list of possibilities weighing the likelihood of some special monk seal need we had overlooked, some invisible disease he might have contracted, or some piece of bone or debris inadvertently stuck in his throat.

"He just isn't interested in eating." Beau sighed. His soaked shorts were a clear sign that he had just had a head-to-head session with the seal.

I knew that Beau and Traci had probably tried every trainer trick in the book to spark the animal's appetite. Conning, ignoring, favorite foods, and downright begging. If one of them was in my office with this news, there was no use in offering suggestions.

When I looked KP2 over from head to toe, his problem was immediately apparent. He was turning into a teenager, with all the requisite hormonal mood swings. I could tell that something was different just from the size of his head. While teenage boys tend to grow feet first, male seals quickly develop blocky heads and chests—those body parts most important in the fight for winning mates. KP2's head looked bigger, at least relative to the rest of his body. Watching Beau pet him was like watching Kobe Bryant palm a basketball.

The seal's vocalizations were another clue that he was getting older. Walking by the side of his pool, I heard KP2 "chugging" to himself underwater. Beginning as a series of steam engine chuffs, KP2's call ended in a long trailing "*yooooowwwwlllooo*." This was a new addition to his vocal repertoire. I was familiar with his raspy-throated "*rrrraughssss*," abrupt "*brahhahahas*," the Three Stooges' "*whup, whup, whups*," and assorted snorts. This new sound vibrated underwater through his throat and chest with nary a bubble appearing. Unlike humans, KP2 didn't need to exhale to create an amazing array of sounds. The low vibrations resonated off the cement walls of his pool and could be felt as well as heard by any passersby. Almost overnight, the seal matured from puppy calls to underwater baritone rumblings.

"It reminds me of echolocating Weddell seals," I commented to Traci as she came out of the food prep kitchen toting a bucket of herring to feed the crooning seal. In Antarctica, while standing on the sea ice we would often feel the eerie vocalizations of the seals through the soles of our Bunny boots as the animals swam below us.

Traci cocked her head toward KP2's pool. "He sounds to me more like Burnyce wailing during mating season," she responded. Traci walked off, juggling the fish bucket to begin KP2's breakfast training session.

Named for Professor Burney LeBoeuf, a retired UCSC elephant seal biologist, Burnyce was enormous. She was part of an acoustic research project at the lab and lived in a pool across the walkway from KP2. Snot perpetually dripped from her bulbous nose. The two seals had many things in common. She, too, had been rescued as a pup and brought to Long Marine Lab. She, too, was inexplicably going blind. Burnyce was also a seal, but of proportions that would outweigh KP2 by nearly three times when he finally reached his full adult mass. At her last weigh-in, Burnyce literally tipped the scales at 1,852 pounds, nearly a metric ton. She was the equivalent of two Harley-Davidson road hogs with sidecars to KP2's Vespa scooter. As her species name implied, Burnyce defined the "elephant" in elephant seal.

Every spring, as the days grew longer and wild elephant seals gathered to mate twenty miles up the coast at Año Nuevo State Reserve, Burnyce would "sing" to them. Singing, of course, is in the ear of the beholder. To the workers in the lab and the surrounding buildings, she was the ritual foghorn that drove the uninformed crazy in its volume and consistency. For days Burnyce's call was a quest for animal sex like no other. The giant seal would haul her massive frame onto the deck of her pool, raise her head skyward, extend her bristly whiskers forward, and sing in a long, low "*ooooooooooooowwwwww.*"

At night the calls increased in intensity. Burnyce howled at the moon, singing along with the lovesick coyotes under the redwoods in the surrounding Santa Cruz Mountains. Nearby ragtag hippies tending the Homeless Garden Project on the vacant lot next to the marine lab smiled in recognition of her calls and the upcoming growing season. When Burnyce sang, spring was in the air.

Only ten feet away, KP2 watched and listened from his sealarium. Neither he nor any of his species would ever have seen or heard anything quite like Burnyce. Without the benefit of hearing the mating calls of his own kind, KP2 was immediately smitten.

To Burnyce he called, "*Yooooowwwwlllooo.*"

"*Oooowwwoooo,*" Burnyce howled back, without giving him the satisfaction of even looking in his direction. Given her great bulk, I suspect that turning to respond to KP2's calls would have been physically impossible, even if the female elephant seal were attracted to her comparatively diminutive suitor.

The language barrier didn't help. Despite the fact that they were cousin species, to Burnyce's and KP2's nearly invisible pinhole ears they were speaking in two completely different seal dialects. KP2's love was doomed from the start.

Although the howling and yodeling of the two seals frayed the nerves of those within earshot, including the cockatoos and Puka and Primo, it diminished quickly without the benefit of consummation. After a week, Burnyce went back to snoozing and oozing snot in her pool, and KP2 amused himself by experimenting with different underwater vocalizations in the phocid version of singing in the shower.

SLIGHTED BY BURNYCE, KP2 soon found ways of flexing his newfound testosterone-infused muscles at the lab.

"Into the water," Traci announced to the hauled-out seal as she entered the sealarium with his breakfast fish bucket in hand.

KP2's normal response had always been an enthusiastic, splashing dive into the pool for playtime. Now the seal opened one eye in a squinting "I don't think so." He then rolled over on his side in the sun.

Feigning boredom or forgetfulness, the seal turned obstinate. This was the turn in behavior that NMFS officials had feared would make KP2 dangerous to people if he'd stayed in the wild. On one level he loved the company of humans, but on another he was still a creature of the wild. Playtime was on his terms, so he began to act. What we didn't know was what would happen if he didn't get his way.

"Water, Hoa!" Traci raised her voice a decibel so the lazy seal would have no doubt as to what she was asking. The increased volume elicited a rollover on the deck with KP2's white belly facing the sun in complete, unmistakable refusal. Traci considered her next move as she did for animal and human alike. She made no distinction when it came to bad behavior. Rude drivers, rude students, and rude animals were treated equally and without latitude. No one was given a break. Traci was the epitome of fairness but also knew that letting her guard down was as good as approval to an offender. She now found herself in a showdown with the teenage KP2.

"Hoa! Water!" Traci insisted. If the seal stalled too long, the trainer would not get mad. She would just walk away, taking the bucket of fish with her.

KP2 waited to budge until it was clear that Traci was on the verge of leaving. Finally, he rolled onto his belly and as leisurely as he could inched to the edge of the pool to slip his body into the water in slow motion. He left his hind flippers hanging in the air just long enough to add a final insult. As soon as he was fully submerged, the seal popped up with an open mouth, looking for a herring handout.

"Oh, no. I'm not going to let you have that one," Traci berated the seal.

KP2 looked expectantly from her to the bucket of fish. When no fish was forthcoming, the seal raised his nose into the air toward Traci and blew out a loud, honking "*sneeeeppphh!*" Seal snot flew in her direction.

On that day the snot shot became KP2's signature display of frustration.

More than once I was the victim of the seal's directed sneeze. It usually occurred when I entered his pool area to measure the water temperature. Up he would pop with his dripping whiskers and nose snuffling on my pant leg while his tail and flippers dangled in the water. He'd look around excitedly for the shine of the stainless steel fish bucket. If he spied the thermometer, he would realize that I was there on science business.

KP2 was polite enough not to splash me as he leaned back into the water. When I bent down to measure the water temperature, however, he'd pop back up in my face with whiskers extended and—"*Sneeeeppphh!*" A snot shot, right as I was inhaling.

"You bugger!" I always fell for his trick. You would have thought that I'd learned that lesson after working with Puka and Primo for so long. The two dolphins had the same habit of exhaling through their blowholes right into your open eyes and mouth if you weren't careful. The smell of rotten fish and dolphin snot lasted for hours.

Wiping my face from the newest attack, I wondered if KP2 had figured out this trick on his own or if he had learned it from his pool neighbors. The placement of the hole in the head was different, but the aim was just as true.

IGNORING THE saltwater-and-mucus spray, Traci maintained a professional upper hand when KP2 was in one of his stubborn moods. To break the cycle, she went back to basic training with the teenage seal. With the thoroughness of a drill sergeant she asked the recalcitrant KP2 to haul out on deck, lie down, and then reenter the water again and again in a series of seal calisthenics.

"Good, Hoa." She encouraged the seal to put more snap into his response. Warming to the game, KP2 would chase after Traci. He'd surf across the water as she ran along the pool deck to the left. Haul out, fish rewards, and splash back into the water. Then the pair would race across to the right side of the pool to repeat the maneuver. Then left, then right. By the end of the session Traci would be panting and the seal shaking with his old puppylike enthusiasm.

BACK IN HAWAII, one of KP2's earliest caretakers, Donna Festa, found another solution to KP2's teenage woes. While conducting beach observations with the Hawaiian Monk Seal Response Team Oahu (HMSRTO), she spied a "real cutie" on Nimitz Beach. NOAA Number 4DF was named Maka'iwi, and she was all that a young male monk seal could ask for. She was sleek, silvery, and, according to the volunteer observers, a real character. She posed for pictures, was attracted to action and fun, and yet oblivious to her many admirers. Some monk seals just brim with personality. KP2 and Maka'iwi shared a natural enjoyment of life that was obvious to all.

On KP2's Facebook page, Donna posted a picture of Maka'iwi with a flipper raised in an "Aloha" greeting instigating a series of "love letters" between the seals. KP2's exchanges with Maka'iwi made folks smile, and it was deemed by his Facebook friends that they should one day have the chance to be together and make beautiful pups.

The Facebook exchanges soon struck a chord with many of KP2's fans back in the islands. The contrast between the lifestyles of KP2 and Maka'iwi was all too evident; try as we might, we could never create the islands in KP2's marine lab sealarium. He was a native-born boy who had now grown into a young adult. The locals wanted him back; he had been gone too long. Thus, through the digital kinship of the social network, a movement to bring KP2 back home to the islands evolved.

A rallying cry in the form of a song was written by Lono, a local Molokai slack-key guitarist. In the lyrics there was no mistaking the connection between the islanders and the young seal. Nor was there any mistaking the seal's power. He was no longer KP2, known just by a government-issued ID; he was Ho'ailona, a sign that could help people renew their connection to the oceans. Lono sang:

Molokai nui a Hina	Great Molokai child of Hina
Home hānau Pule O'o	Birthplace of the powerful prayers
'āina Kaiakea me	Land of Kaiakea and
Kuapaka'a	Kuapaka'a
Please, oh please, bring	Please, oh please, bring
Ho'ailona home!	Ho'ailona home!

I loved Lono's song for the seal, and his message. But the question was where? Where do you put a maturing, nearly blind seal who was capable of loving people to death and would naively face an angry fisherman's gun with an enthusiastic whiskered greeting?

17.

A Roof Above

⁓

Taking care of a seal is not easy. A bathtub will not do. Besides the mountain of permits and governmental inspections regarding housing, there is the cost of fish, vitamins, personal trainers, and veterinary care for an animal that can live as long as thirty years. You have to be a little crazy and in it for the long haul, especially for an endangered seal. Add in the cost of medications for his eyes, and it was no wonder that so few facilities were willing to take the risk of housing KP2.

As much as the Hawaiians wanted the seal back, they could not agree on where he would go. Walter Ritte maintained his original stance. He wanted the animal returned to Molokai to live in an abandoned fishpond where the children could continue to interact with him. Donna Festa and members of HMSRTO opted for Sea Life Park on Oahu, an older aquarium from the 1960s where a docent program could be established. NMFS in Washington saw the Waikiki Aquarium as the ideal placement facility due to its location and research connections with the University of Hawaii.

Traci, Beau, and I wanted only one thing. Wherever KP2 went, he could not be ignored. We knew that, isolated from people in a feeder pool, the seal would wither away.

As discussions about where to place the seal in Hawaii wore on, we faced a bigger, more immediate housing problem of our own. In the annual tug-of-war between winter and summer that was known as spring in Santa Cruz, one last storm crashed in from the Arctic to wreak havoc at the lab. One wild night water pounded on KP2's sealarium roof, challenging the plastic-and-rope canopy that Traci and Nate had so carefully constructed. Throughout the night, the weather station on top of the lab strained under winds clocking a continuous gale force of thirty-five miles per hour and the occasional gust to sixty miles per hour. Inside the sealarium the plastic shook and snapped with the ferocity of a sail that had broken from its rigging. The trainers and I barely slept through the high winds and hailstorm, worrying about the roof and KP2's welfare. Meanwhile, the seal was busy dealing with the storm on his own.

The following morning was cold and crystal blue. After checking on KP2, the trainers prepared his fish breakfast while I calibrated the instruments for our next metabolic trial with him. Unexpectedly, Nate, the facilities maintenance man for the university, came by and asked, "So did you see?"

I didn't know what he was talking about.

"The roof," he explained. "Interesting hole."

Beau, Traci, and I ran out to inspect the seal's pool. KP2 came surfing over, acting as if nothing unusual had happened. At first glance the structure seemed to have weathered the storm with little more than some sagging in the plastic roof where rainwater had collected. Then we saw the ragged tear. Rather than a simple hole, the plastic was shredded in the center and the ropes that formed the backbone of the roof were stretched beyond repair. The entire structure was in danger of collapsing inward.

Traci looked from the hole to KP2. The damage was more than eight feet above his head, but instinctively she knew that somehow the seal was behind the destruction.

Nate brought out a twelve-foot-high orange step ladder and placed it in the pool as far into the center as he could. Stepping gingerly from the edge, he balanced himself for a closer look at the hole. KP2 slunk away to the far side of his pool, acting suspiciously like a bad dog who knew his crime was about to be discovered.

Pulling gently down on the springy plastic, the maintenance man realized that the entire roof easily sagged to within a couple of feet of the pool surface. "You know, I do think these are bite marks," he said slowly.

The evidence was clear. With rainwater accumulating on the roof overnight, the supporting ropes had given way, allowing the plastic to sag within jumping range of the monk seal. KP2 had obviously spent the night playing, leaping and splashing to snap at the plastic and ropes. He had bitten through in several locations, allowing the rainwater to shower over his face like a sieve. By morning the roof and plastic had sprung back into place, almost allowing the seal to get away with his misdeeds.

"Oh, that would be just like a seal!" Traci glared at KP2, who would not meet her accusing eyes.

Moving the ladder to the deepest part of the pool, Nate began to remove the damaged rope and plastic. With the rest of us out of sight,

KP2 overcame his guilt, and curiosity soon got the better of him. He circled the ladder and then tried to work his body in between the submerged rungs as if he were navigating a coral reef. He twisted and poked his blocky head into each rung. Finally, he greeted the man atop the ladder by hurling his body onto the steps. When that didn't work, he hugged the legs of the ladder in an attempt to climb up. It was a remarkable, agile performance considering he didn't have the benefit of hind feet.

"Whoa, oh! Guys?" Nate called to anyone within earshot. "Uh, guys, I need some help here!"

KP2 knew that he had Nate cornered and was unrelenting in his pursuit. The rambunctious seal continued to climb up, causing the ladder to wobble precariously and bring the frantic human to him. But the frightened man counterbalanced and avoided the seal by moving even higher. Diving to the pool bottom, KP2 began biting the protective plastic feet of the ladder.

By the time Beau and I arrived, Nate was hanging awkwardly from the ropes with his legs entangled in the ladder and KP2 circling like a shark.

"So how's it going up there?" the trainer asked cheerily.

"He was after me! That seal is crazy!"

"Hoa?" Beau queried the seal. Together we distracted KP2 long enough for the maintenance man to escape. However, in trying to remove the step ladder we quickly discovered that it was not the human the seal was attracted to—it was the ladder. KP2 dove and wrapped his body around the legs of the ladder. Thus began a tug-of-war, with the seal using his body weight to pull the ladder back into the water while we slipped across the deck trying to pull in the opposite direction.

"You'd think this was his pink boogie board," I complained as I struggled with the seal. KP2's response was so focused that the best we could do was drag him up on deck entwined in the lower rungs of the ladder. Only later, in a conversation with KP2's Hawaiian caretakers, did I piece together KP2's unusual attraction to stepladders. For the first seven months of his life at the Kewalo Research Facility, ladders had been the seal's only connection to people, food, and nurturing. When stepladders arrived in his small pool, humans had joined him. When stepladders left, so did all their attention. KP2 was determined not to let another ladder leave without a fierce struggle, a habit he never outgrew at the marine lab.

EVENTUALLY, WE WERE ABLE to repair the damage to KP2's sealarium roof, but not without adding one more in a never-ending trail of bills. From the beginning, taking care of KP2 and conducting our science was a risky cat-and-mouse game between me and the university accountants. I was growing increasingly uneasy with my inability to keep in front of the costs. The dismal economic climate in California did not help.

Unprecedented budget cuts were tearing the heart out of the University of California and threatened to shut our campus down. "Governator" Arnold Schwarzenegger and soon thereafter Governor Jerry Brown attacked the state deficit by squeezing the university. Each $500 million hit created a deeper sense of doom at the lab, and I began to dread e-mails from UCSC chancellor George Blumenthal, already a basset hound of a man, who grew even more hangdog with each cut to our campus.

Salary reductions and furloughs, hikes in student fees, and staff layoffs crushed the spirit out of our tight-knit academic community in the redwoods. Student anger mounted and spilled over into sit-ins and protests that destroyed part of the UCSC admin building. Fortunately, some students took a different tack.

With so few paying jobs available, students arrived on my doorstep volunteering to help care for KP2, Puka and Primo, Wick, Morgan, and Taylor, and the birds. To my amazement and relief they willingly dedicated over sixteen hours a week working for free at a time when they needed money most. Without this corps of volunteers to clean pools, thaw fish, and train and feed animals, we surely would have gone under.

The volunteers from my lab were easily identifiable on campus. Their clothes were bleach stained from cleaning pools and there were fish scales lodged beneath their fingernails. The difference was also apparent on their faces. In contrast to so many of their classmates, they appeared happy. It all stemmed from living their passion, and sharing it with a cohort of like-thinking individuals. In the company of KP2 they had become 'ohana and dreamed of making a difference in an increasingly unstable, self-absorbed world.

I let the students continue to live their dream, never mentioning the bills that consumed my waking hours and then some. Neither the student volunteers nor the trainers knew how close we came to shutting down. A major blow came when a grant proposal to support our research with KP2 was declined by the Marine Mammal Commission. Governmental funds for the Hawaiian monk seal program had been cut in Washington; the trickle-down effect meant no funds for outside researchers. The news was devastating. We were already barely scraping by.

A second event following soon thereafter brought me to my knees.

Budget cuts to the university instigated a scavenger hunt among campus administrators who scoured financial reports and bills. With the intensity of starving coyotes, they searched for any way to save money. Campus recreational programs were downsized, teaching assistantships denied, staff retired, and faculty pressed to breaking. Finally, as university accountants nipped away at every possible financial corner, they found KP2's and my hiding place. I had hoped that the budget crisis and furloughs would have kept them confused a bit longer.

"You owe the university $30,000 for heating salt water in the animal pools," Randolph's boss told me.

With a massive bill in hand, my heart sank. I was finally caught.

18.

Mother's Day

⁓

In an attempt to raise at least a little bit of cash for taking care of KP2, I decided to host a 10K race. Local children and adults were invited to the lab, with all proceeds from the race entry fees going to support the care of the young monk seal.

Two triathlete friends, Penni Bengston and Mary Zavanelli, immediately took charge. They created monk seal posters and a race Web site without the benefit of ever having met the tropical seal. They also decided to host the race on Mother's Day.

"I really don't—" I stammered when they picked the second Sunday in May.

"It's perfect!" Penni countered, cutting me off. "KP2 is an orphan. What better way to celebrate the day with him!" I reluctantly agreed.

As uneasy as I felt about the choice of Mother's Day, I added one more detail. I wanted an authentic Hawaiian chanter to bless the runners before the race. There would be no mistaking the island connection to our venue.

MOTHER'S DAY ARRIVED with the rising sun casting a rosy glow over the lab. As Penni, Mary, and a crew of volunteers set up the racecourse, there was little for me to do until the athletes arrived. So I sought the solitude of KP2's enclosure.

With all attention at the lab focused on race preparations, KP2 was bored. He had abandoned his usual sleeping spot and was busily nosing the bottom of a pool gate when I entered the sealarium. He never heard the opening and closing creaks of the door, or perceived the motion of my feet on the deck. I sat down quietly in the corner watching him.

The young seal had no perception of mothers or Mother's Day. This was not surprising, since he had not had much of a mother. As mothers go, phocid seals are a schizophrenic lot, blending a self-sacrificing character with sexual promiscuity and child abandonment. KP2's mother had taken this phocid personality to an extreme when she attacked him.

For seals, the period of maternal bonding from birth to weaning is a blink in time compared with that of dolphins and humans, who spend years nursing their young. Arctic hooded seals have the shortest nursing period of any terrestrial or marine mammal, a mere four days of interaction between moms and pups. Even rat mothers nurse longer. The other end of the seal spectrum is Russia's rotund Baikal seal moms, which nurse their offspring for two months. I always found it interesting that these Russian seals were the smallest of all true seals but provided the longest nutritional head start for their offspring.

With a six-week nursing period, however, the typical Hawaiian monk seal mother was not far behind.

When nursing, seal moms remain with their pups and forgo any other activity, including eating. Such focused devotion literally deflates these seal moms, as blubbery reserves are turned into milk to feed their ravenous pups. And this is no ordinary milk. The milk of a seal mother is so rich in fat that it is closer to cream. Human milk and cow milk are approximately 4–5 percent fat. By comparison, the milk of phocid seals is ten times richer, reaching a remarkable 65 percent fat in some species, almost twice the fat of whipping cream. Creating such a fatty milk diet takes a lot out of these marine mammal moms. By the end of a nursing period, the abdomens of the moms will have caved inward, exposing the sharp outlines of their hips and shoulders.

Rather than having to lose pregnancy pounds like human mothers, seal moms must return to the sea and feed heartily to regain their original streamlined profile. Meanwhile, their pups will have taken on the shape of beach balls with flippers. These rotund youngsters lazily digest the milk they have stored in enormous blubber layers for weeks, much like chicks use stored yolk to sustain their growth until they hatch. Eventually, hunger drives the young "weaners" to sea, where they learn to hunt for fish on their own. Through trial and error the seal pups begin to catch their meals.

Seal mothers never know of their offspring's accomplishments or growth. They are long gone, seeking food and mates for next year's pup.

Yet there is a unique facet to Hawaiian monk seal mothering that follows in the island matriarchal custom. Unlike other seal species, Hawaiian monk seals are exceptional in their altruistic behavior. Ignoring the drain on their own milk resources, they unreservedly play "auntie" to the pups of other females. A hungry mouth, regardless of whose pup it belongs to, will be offered an open teat. In this way abandoned pups avoid starvation.

Unfortunately for KP2, there were no other lactating females on the beach when his mother left. Otherwise his life would have taken a completely different trajectory. Motherhood had not been a part of his fate.

So, too, Mother's Day left me empty.

I WILL NEVER KNOW what prompted Austin to walk across the neighboring mountain ridge on Mother's Day two years before. When I returned home from the lab, my beloved corgi was nowhere to be found. The rush of panic was immediate. Austin had never left the property. I walked the mowed grassy acres and local roads for hours into the night, calling his name in a voice shifting in pitch from deeply angry to thinly fearful. Throughout the evening his metal dinner bowl remained untouched on the front deck and by morning the blue jays had started a raucous war over his kibble.

There was no time to continue my desperate search for the corgi. Students were waiting at school, and I had to put my emotions in check in order to teach. I'd retrieve them later when there was time to turn off the scientist. As I walked to my biology class I designed lost-dog reward posters in my head and planned an afternoon of searching the local woods.

Several hours later, when I finally returned to my office, a message from Austin's veterinarian was on my phone. I hesitated when I saw the blinking message light and then quietly shut my office door. I wanted privacy for a phone call that I knew would confirm what I already felt in my heart.

"A dog matching Austin's description has been found . . ." The veterinarian paused. "He was wearing a rabies vaccination tag from our office on his collar . . ." He stalled again and couldn't get the words out.

"Where?" I choked. My pulse was racing. I knew.

"Your neighbor's."

My world turned colorless. Nothing seemed real as I drove home in a numbed stupor. Unconnected flashes of reality would leave permanent scars. A drive up a mountain road, a soaked blanket, a white paw, and a heavy wet body. Dirt and a backyard grave. Sobbing and darkness.

"Stupid, stupid!" I grieved angrily and endlessly. I sat outside by the overturned dirt of the grave site, my legs too numb to move.

I didn't know how to do this; for the first time in my life I didn't know how to act the part. What does a grieving scientist look like? There seemed no relief from the ache. Searching for any respite, I tried to rationalize myself out of the despair. "You know, through MRI scans scientists have discovered that intense emotional pain and physical pain activate the same neurological pathways in the brain. Maybe that's why you can't stop crying." Then I broke down again. Science would not rescue me; I could not think my way out of the situation this time.

For days I repeated the scenario over and over again in my head. Each time I could smell the decay of the redwood forest and hear the gurgling of the small creek next to the house. Had Austin detected the tracks of a coyote and followed its trail up and over the mountain ridge? Did he see sunlight in the clearing of the neighbor's house? Had sounds attracted him? No one was home when he'd entered their yard.

Unbeknownst to the small dog, his personal demon lay in wait. Austin's path was blocked by the one monster we had so often battled together in Hawaii: water. Irrationally, paradoxically, he tackled his nemesis as instinct drove him to do. Austin plunged headfirst into the neighbor's swimming pool. He pumped his stubby legs and swam frantically to reach the other side as he had on every ocean kayak trip and hike to Sacred Falls with me.

Here the tape in my head runs out. Did he ever reach the side of the pool? Were his legs too short to allow himself to pull his body over the edge? Did he swim to the bottom and inhale water on the way or did his legs simply grow progressively weaker as minutes passed into hours? Ultimately, the grief was not in how it happened; it was in my failure to be there to rescue him this time. In my mind I had killed the only animal who could see into my soul.

But there was much more sorrow in my tears than for the loss of this dog. Here was my payback for entering the man's world of science. This was the one sacrifice they could never know. For my entire career I had been able to laugh in the face of glass ceilings, sexual harassment and innuendos, unequal pay, and discrimination. They were nothing compared to the ultimate sacrifice, that of motherhood. In dedicating my life to the creatures of the wild, home and family were as ephemeral as the clouds I had experienced over the Namibian desert or the tidal pools I'd swum in along the Kona coast. Austin was not just a reader of my soul, he was the child I would never have. He was my *'ohana*. And now he was gone forever.

AT LAST, KP2 SAW ME sitting with my knees hugged to my chest, brooding in the corner of the sealarium. He surfed across his pool at full speed to greet me. Stopping at the pool edge, he gauged my reaction. When I didn't move, he used his fore flippers to slowly haul his body onto the deck. He inchwormed his way in my direction, all the time reading my mood from my downcast eyes and body movements. He showed no hesitation as he sniffed my pant leg. Then he laid his blocky head on my foot and shut his eyes.

My first inclination was to back away as I always did. Instead, I looked down at the sleeping seal and then placed a hand on his back. I began to stroke his fur. "I'm sorry. I don't have any fish to give you," I apologized, thinking he would move off once he realized that no reward was forthcoming. But KP2 didn't look up at the sound of my voice. He was content to remain by my side. After several minutes I added, "Looks like Mother's Day isn't all that great for you, either, is it?"

We sat silently together, with me lost in thoughts of sacrifice and the impossibility of my lofty conservation goals. As time passed, I became more depressed as I added up the hurdles we faced. I watched KP2's chest rise and fall with each deep breath. I could feel and see each heartbeat reverberate along his body. In petting his head and then his body, I was surprised at the warmth. Wet dolphins and sea lions always felt clammy and cold. KP2 was like a warm puppy, comforting to the touch.

"It is hopeless, isn't it?" I tested the seal. I sighed, thinking I had finally had enough. It was time to stop deluding myself about saving endangered species.

I bent down to look KP2 square in the face to inform him of my decision. I was done.

The seal's brown eyes opened as he felt my breath near him. Our faces were so close I could see the dull edges of the gray spots outlining his cataracts. The seal studied my face, although I wondered if he could even see me through his damaged eyes.

Before I could say a word, KP2 snorted loudly without warning. He sprayed my eyes and face with snot.

"You bugger! I know you did that on purpose!" I cried as I tried to wipe away the smell of seawater and rotting fish. I would smell his fishy breath on me for the rest of the day, no matter how many times I washed my face.

KP2 rolled back into the water with a splash and refused to surface. Instead he went back to playing with the gate bottom and vocalizing to Puka and Primo on the other side. It was his form of the cold shoulder.

I was ready to slap the water and yell "Get back here!" when I just stopped. KP2 resurfaced, sat up in the water, looked at me, and then saluted. The trainers had thought it would be a cute behavior for the seal to have in his repertoire—all sea lions knew the salute command. The problem was that KP2, like all phocid seals, had short front flippers compared with seal lions. He could not reach his forehead. The best he could do was wrap his stunted flipper across his muzzle. KP2 looked like he was holding his nose instead of saluting.

The seal knew exactly what he was doing. I started to laugh until I was crying and then began laughing again. I couldn't stop, and I realized that KP2 knew me better than I knew myself. It had been a long time since I had experienced such a sense of closeness with an animal. Not since Austin.

It was also the first time in years that I had laughed on Mother's Day. The young monk seal that had been attacked and abandoned by his own mother had never spent one moment looking backward. He had been kicked and caged, transported and trucked. He had lived in pools and pens, experienced the wild ocean of Hawaii and the California winter. Life was all about the adventures in front of him, and he always seemed to make the best of it.

"Okay, you win," I told the seal. "I get it. Stop feeling sorry for myself. No one said saving your species would be easy."

I never told Beau or Traci about that morning with KP2. I'm sure they would have disapproved of my poor training techniques. But that shared moment with KP2 spurred me on to help the Hawaiian monk seal using every resource within my reach and more. We would find him a new Hawaiian home and I would help him get his environmental message out. I had no choice. I had finally found the one wild animal who could read me.

COVERED IN FRESH SEAL SNOT, I noticed that the beginning of race day was as picturesque as any in Santa Cruz in May. Even to my bleary eyes the blue skies and light breezes over the marine lab looked inviting.

Hundreds of smiling racers soon arrived with the bustling nervousness of thoroughbred horses awaiting the derby. The driveway outside of KP2's sealarium buzzed with the adrenaline of competitors. As race time approached, Mehana, the Hawaiian chanter I had hired, appeared in a dark dress, flower lei, and headband of braided beige kukui nuts. She gathered the racers around her. Children and adults in racing gear stood by the blue whale skeleton mascot of the lab, mesmerized by the woman's chant as she called to the land, the oceans, and the sky. She spoke of the islands, its people and its animals, and a special monk seal blessed as Hoʻailona.

Mehana continued to chant as clouds began to build up over the lab. With each wailing verse the sky drew darker and more ominous over us. The winds seemed to respond to her calls, threatening to blow over the balloon arch at the starting line. Then, with a final sweeping outreach of her hands, Mehana released the runners onto the racecourse, and the clouds released their contents upon us.

Some attributed the unusual late spring rainfall to the tail end of the El Niño weather pattern along the California coast. Although scientifically valid, I considered the capricious timing the handiwork of the mischievous Madame Pele, whose wrath the locals of Oahu had warned me about when I'd first met KP2 at the Waikiki Aquarium.

With each soaked runner that passed the starting line, I began to wonder if Mehana had awakened the curse of Stolen Child in invoking the spirit of the islands with her early morning chant. Perhaps this was the final revenge. Or maybe it was a reminder from Hawaiians that KP2 was a temporary gift to the mainland. His home was back on the islands.

As the runners crossed the finish line and visited with KP2 and the dolphins, I vowed to contact Amy Sloan at NMFS to begin the process of returning the seal to his native land. Although I'm not prone to superstition, I did note that the clouds dissipated and the sun broke through the moment I made this commitment. For the rest of the day, KP2 entertained his visitors by surfing in the glow of the California sun. He experienced the first Mother's Day of his life.

I LEFT THE MARINE LAB feeling a sense of warmth about Mother's Day that had long escaped me. Irrespective of my scientific training and the untimely loss of Austin, I still harbor the romantic notion that all mothers, including the runaway RK22, are inherently saintly. Maybe it is my inexperience with their job that makes me think this way. It was KP2's misfortune to have an inept mother. He was never afforded a drop of her milk, but I always wondered if abandonment had been the real motive behind RK22's actions.

Life is not all paradise in the world of Hawaiian monk seals. With the marked decline in numbers came a skewing of the sex ratio in five major breeding areas in the Northwestern Hawaiian Islands. The most vulnerable segments of the population, the very young and the female, were the first to disappear. Now, instead of the requisite one male for one female to create a pup, there were at least two males for every female. Beaches had become as competitive as any Saturday night bar, where the entrance of an unattached female invariably instigated a brawl among cruising males.

The overenthusiastic ardor of Hawaiian monk seal males has had devastating effects on the population. Adult females and immature seals of both sexes are attacked by roving gangs of four to fourteen testosterone-high males, all seeking mates. The resulting melee can last for hours, with the victims either severely injured or killed outright. If injured, the incapacitated female or immature seal is likely finished off by tiger sharks patrolling the waters. These "mobbing" incidents only exacerbate the population problem. By selectively increasing the mortality of females and young animals, male gangs have sent the male-female sex ratio and the Hawaiian monk seal population into a death spiral.

I began to think that maybe the aberrant sexual behavior of male monk seals was at the root of RK22's peculiar maternal behavior. Several bullying males had been observed in the area where KP2 was born, with one aggressive male even attacking RK22's newborn pup. Observers reported that her response had been to attack KP2 and then swim off nonchalantly with the boys, abandoning her pup. That was one interpretation.

What if RK22 had another intention? What if, after years of harassment from mobbing gangs of male seals, monk seal moms had developed a new survival strategy? In abandoning her pup, KP2's mom had lured the most aggressive males away from her offspring. In choosing a male to mate with rather than waiting for more aggressive attackers or a murdering group, she could guarantee that there would be another pup in the future and prevent her own injury or death. In a world turned upside down by violence, the case could be made that RK22 had made the supreme motherly sacrifice for the future of the species.

On this Mother's Day, I liked to think that was the way it happened.

19.

The Hand of Man

—

My resolve to return KP2 to the islands was stalled by equally forceful conflicting opinions developing across the Pacific Ocean and on the opposite coast of North America. In Hawaii the celebration of the seal's second birthday on May 1 brought out island emotions on his Facebook page. Traci had constructed a sandbox for the seal as a birthday present. Her intention had been to cheer him up by reminding him of the beaches in Hawaii. KP2 immediately took to the sand, rolling in it and burrowing his head to the bottom. The pictures delighted people when they saw the "sugarcoated" seal. More than a hundred *Hau'oli lā hānau* (Happy Birthday) messages arrived, mixed with pleas for him to come back home to Hawaii. The sandbox, originally intended as a present, highlighted the seal's wild island origins. Across the miles I could feel the groundswell of pressure growing from the islands as locals, who'd originally been promised by NMFS officials that their native son would return within a year, began to wonder if they had been duped.

In response, Amy Sloan and Jennifer Skidmore from NMFS began to make inquiries as to what it would take to move KP2 back to the islands. They had much more than a small vested interest in the transfer. Two years earlier they had put their reputations on the line to move KP2 to my mainland lab. The ultimate proof of their idea that you could bring an endangered species to the scientist rather than just bring the scientist to the endangered species required that KP2 safely return to Hawaii.

Amy began a series of conference calls involving veterinarians who were needed to ensure KP2's health, aquarium facilities that could serve as a potential new home for the returning seal, assorted Hawaiian Monk Seal Recovery Team members, and NMFS officials. Almost immediately plans for the seal's return hit a roadblock.

"It's good that you weren't there." Amy was talking as quietly as possible and fighting back tears when she called me after one of her conversations. Her conference calls had uncovered the disappointing truth.

"You mean to tell me that *nothing* has happened?" I asked incredulously. I had expected that at least some plans and construction would have taken place in preparation for housing the orphaned seal when he returned to Hawaii. "We've had KP2 for fifteen months and nothing has been resolved in the islands? He was supposed to leave after a year. How can the folks there not understand? Where is he to go?" I asked too many questions that I knew had no answers.

"You know that things move slowly in Hawaii . . ." Amy tried to rationalize the fact that nothing had changed since KP2 had made his transoceanic flight. Pools at the only two aquariums able to house monk seals in Hawaii, the Waikiki Aquarium and Sea Life Park on Oahu, were still full to capacity. No other facility or pool was forthcoming, nor apparently had anyone really looked into alternatives.

Jennifer was listening on the line. Her job at NMFS was to find homes for the injured, sick, stranded, and otherwise nonreleasable marine mammals. I waited for her to weigh in on our conversation. Jennifer finally noted, with stark frankness, "In cases like this where there is no place for the animal . . . well, they are euthanized."

My head began to spin. How could this be happening? I knew that Jennifer wasn't proposing killing KP2; she was merely reporting the gravity of the situation.

"That is insane!" I protested. I saw the faces of Beau, Traci, and all the student volunteers. What have I done, I thought. These people believed in me and the dream of changing their world. They had complete trust in our ability to save a species, beginning with this one seal. I had promised them it would happen if they had the courage to take the risk. Every day I saw the hope in their eyes and the faith in their actions. I could never tell them what I had just been told by Jennifer.

I also saw my computer screen filled with birthday greetings for the seal and the wishes from the children of Molokai who'd once befriended him. I recalled all his aunties and uncles who had nursed the tiny newborn pup two years earlier. Mostly I saw a young, comical, high-spirited seal whose only crime was loving people too much. It just couldn't end like this. I was heartsick.

"I'm sorry. I have to go . . . teach," I muttered, making up an excuse to end the call. I needed to think. KP2 and I were running out of time.

"Don't worry. We're going to fight this!" Amy said quickly before I hung up.

Hopping onto my road bike, I pedaled furiously along the coast to clear my head. How had things gone so wrong? I chastised myself for letting my emotions about KP2 and the plight of Hawaiian monk seals get the better of me.

The situation in Hawaii had become more explosive in the months since KP2 had left. On the heels of the monk seal shootings, Lieutenant Governor James "Duke" Aiona signed Senate Bill 2441, making it a felony under Hawaii law to harass, harm, or kill any endangered or threatened species. Punishment for injuring a monk seal now included a fine of up to $50,000 and five years in prison. While this should have been good news for the seals, some fishermen were incensed. "What if a monk seal that is stealing fish from my lines inadvertently becomes entangled?" they demanded. "What then?"

A simultaneous rejoinder by a local fishermen's group declared that monk seals were not afforded the status of 'aumakua. Rather, the marine mammals were nothing more than an "invasive species" to the Hawaiian Islands. The seals had come from the Caribbean and were no more native than the Norwegian rats that had jumped from ships or the mongoose that had been introduced by the government to kill the invading rats.

DESPITE MY ANGER, I had to admit that, as incredible as it sounded, there was some evolutionary truth to the fishermen's claim that monk seals were an invasive species to the islands. It all depended in how far back in history you looked.

The birth of the monk seal lineage, the Monochinae, occurred eleven to fourteen million years ago along the rugged coastline of Turkey and Greece when the body of water was called the Tethys Sea. These ancestral monk seals, later named the Mediterranean monk seal

(*Monachus monachus*), far predated human civilization. They arrived in the shadow of the age of mastodons as the massive land mammals were sinking into extinction. At the time, the oceans were much warmer than today and were filled with a bounty of food that the seals thrived on.

The monk seals were not alone. Enormous marine competitors also vied for the riches of the sea to satisfy ravenous appetites. Giant, toothy predators had been lurking in the oceans for over four hundred million years before the comparatively diminutive seals dipped a flipper into the water. These competitors were of a size and ferocity far beyond anything that has ever been witnessed by modern man.

One threat was the megalodon, aptly named "big tooth" by the Greeks. An enormous shark fifty-two feet long with a gaping jaw that could easily swallow adult monk seals whole, the megalodon ruled the early seas for twenty-seven million years. Rows of teeth exceeding seven inches in length guaranteed the shark's position at the top of the food chain.

Equally dangerous were the ancestors of the snakelike, jagged-toothed basilosaurus. As the first giant whale, basilosaurus swam the Tethys Sea forty-five to thirty-six million years ago; it was the embodiment of "sea monster." Averaging sixty feet in length, this whale has been surpassed in size only by the blue whale, and was the largest animal living at that time. The metabolic demands of the lithe and fast basilosaurus pummeled the smaller animals of the seas. Fossilized remains of these whales have shown evidence of preying not only on fish, but on every other ocean neighbor, including small sharks, sea cows (the extinct herbivorous relative of dugongs and manatees), marine turtles, and other whale and dolphin species. Although predating the arrival of the Monochinae by twenty million years, the killer whale–

like relatives of basilosaurus retained the habit of cruising the shallow waters of the Tethys Sea, making the sluggish, rotund monk seal a tasty target.

The ancestors of KP2 had one advantage over their oceanic competitors, however. Unlike the gigantic sharks and whales, the seals retained their ability to return to land. Lumbering awkwardly on stubby front flippers, the seals wormed their way onto the rocky shorelines and hid in coastal caves. It was here that they were eventually discovered by a new, cunning predator: man.

As the earth's oceans grew, so did the range of the monk seal. Over the course of several million years, the seals spread throughout the entire Mediterranean Basin and eastern North Atlantic. Fossil and DNA evidence suggest that this seal lineage maintained its peculiar, un-seallike affinity for warm waters and eventually split into two sister taxa, the Caribbean (*Monachus tropicalis*) and Hawaiian (*Monachus schauinslandi*) monk seals.

The seals' incredible transglobal journey took place in two steps. First, they traveled east from the Mediterranean Sea to Caribbean waters via the North Equatorial Current. In a second excursion, the seals swam from the Caribbean to the Hawaiian Islands through the ancient Central American Seaway that once connected the Atlantic and Pacific Oceans. When the seaway closed 3.5 million years ago, Hawaiian monk seals residing in the Pacific Ocean were effectively isolated from their sister lines left on the Atlantic side of the world.

It is difficult to say exactly how and when monk seals arrived in Hawaii. Naysayers have suggested that it was impossible for the animals to swim such great distances to the remote midoceanic islands. Yet humans clearly managed this feat when early Polynesians traveled thousands of miles in dugout canoes from as far away as the Marquesas

Islands. The first human footprint in the soft sands of Hawaii is believed to have occurred between AD 300 and 800, when intrepid Polynesians finally landed in the islands. Molecular, genetic, and anatomical fossil evidence suggests that monk seals were there to greet the humans, having arrived four to eleven million years before man ever paddled into the blue waters. Another millennium in posthuman settlement would pass before Kamehameha the Great, the first king of Hawaii, would unify the islands into the Hawaiian Island group and give the silver-and-white native seals their name.

Based on this history, the protesting fishermen were right. Hawaiian monk seals could be considered invasive. Their ancestors had traveled across the globe and two oceans to reach the islands. However, the historical and scientific evidence also shows that it was man who had invaded the seal's tropical playground, when humans eventually navigated their way to paradise millions of years later.

REGARDLESS OF WHO ARRIVED FIRST, Hawaiian monk seals, like their Mediterranean and Caribbean sisters, experienced a rocky relationship with the men of the sea. It had been that way since the beginning of history.

Ancient monk seals bore witness to the evolution of fishing and the exploitation of the oceans by humans. Along the coastline of the Mediterranean Sea, man represented competitor and predator to the warm-water seals. Along with mussels, fish, and dolphins, monk seals were dismembered, defleshed, and consumed by Neanderthals living in sea caves along the southeast side of Gibraltar Rock. Hunters from later civilizations, particularly the Roman Empire, hunted the monk seal for clothing and medicines. Today we might laugh at some of the seal "cures." Renowned for their ability to sleep soundly, monk seals were believed to possess the ability to cure insomnia, which could be accomplished by placing the right flipper of a monk seal beneath your pillow.

Monk seals dispersing into the West Indies were exploited for meat, skin, and the oil in their blubber. Discovered during Christopher Columbus's second voyage in 1494, the Caribbean monk seal became a source of fresh meat for starving crewmen making transoceanic voyages. Three hundred years later European colonization brought further pressure on the tropical seals as their blubber was processed into cooking and lamp oils, lubricants, and coatings for the bottoms of boats. Processing plants skinned the seal carcasses and turned the exotic seal pelts into everyday household items ranging from trunk linings and articles of clothing to leather straps and bags.

Such unrelenting exploitation could not be sustained by the population, and the catlike smile of the Caribbean monk seal faded into extinction sometime during the 1950s. For decades scientists searched unsuccessfully for the seal. On June 6, 2008, one month after KP2's rocky birth, *Monachus tropicalis* was declared officially extinct.

To date, the Caribbean monk seal remains the only known seal species to have been forced into extinction by the hand of man. The two remaining sister lineages, the Mediterranean monk seal and the Hawaiian monk seal, are teetering on extinction's edge for the same reason. The former, with only four hundred individuals residing in the waters of the Mediterranean and Black seas, is classified as the most endangered pinniped in the world. The third and last sister comprises eleven hundred individuals clinging to the islands of Hawaii. Of the three monk seal lineages that once cruised the warm oceans of the world, the Hawaiian sister, KP2's remnant family, represents the greatest hope for survival.

THE HAND OF MAN

. . .

STORIES IN THE FORM of spoken word, theater, and art revealed the oceanic soul of man and his relationship with the animals of the sea long before the discovery of DNA and the age of science and books. This tradition, too, told of a tumultuous relationship between seals and man. At a time when the borders of Turkey were in dispute, neighboring Greece was a sea empire whose culture was infused with images and stories of oceanic gods, fishermen, dolphins, fish, and other sea creatures. According to Greek mythology Poseidon was the ruler of the seas and all its watery creatures. He was both a creator and destroyer, using the animals of the seas to suit his whims. Greek art is filled with romantic images of Poseidon with trident in hand and hair flowing as he drives a chariot of hippocamps (sea monsters that were half horse and half fish) up from the depths.

While responsible for creating the most beautiful animal on earth—the horse—Poseidon in some ancient accounts is also credited with a few animal "mistakes," not the least of which were the hippopotamus and the seal. To take care of these oddities, Poseidon's son Proteus became the herder of seals. If modern DNA analyses are correct, the members of Proteus's flock could only have been the Mediterranean monk seal, the earliest ancestors of KP2.

Proteus was the prophetic "Old Man of the Sea" who spent his days tending Poseidon's oceanic flock. The description of his daily habits in Homer's *Odyssey* foretold what we now know about the diving activities of modern monk seals, their propensity for hiding in caves, and their distinctive oceanic odor:

When the sun in its course has reached midsky,
the sage old sea-god leaves his ocean . . .

Once ashore, he lies down to sleep under the arching caves,
and around him is a throng of seals . . .
They too have come up through the gray waters,
and they too lie down to sleep,
smelling rankly of the deep brine below.

Interestingly, Greek mythology refers to the necessity of a seal herder—that monk seal numbers were once so great that a shepherd was needed to control them. Today, a modern Proteus has risen from the same seas, not to control their great numbers but to protect the remnants of the ancient Mediterranean monk seal race. Like the mythical Proteus, today's guardians are intimately familiar with the habits of the marine mammals and the waters in which they reside. In an inspiring change of heart, the newest Proteus is a group of Turkish fishermen who have come together to rescue one of their wiliest competitors.

Their efforts to protect Mediterranean monk seals originated in 1991 on the landlocked campus of Middle East Technical University more than one hundred miles from the Black Sea and several hundred miles from the Mediterranean Sea. Students from the Underwater Research Society's Mediterranean Seal Research Group teamed with local fishermen and the World Wildlife Fund to form fishermen's cooperatives and establish no-fishing zones that would protect the coastal caves where monk seals hid, bred, and nursed their pups. They created an intensive policing program to prevent illegal fishing activities and have been carefully monitoring both the population status of monk seals and fish stocks along the outer Gulf of Izmir in the Aegean Sea and the Cilician Basin in the Mediterranean Sea.

Today, over twenty-nine conservation groups, spread across countries bordering the Mediterranean Sea and boasting more members than there are monk seals, are working under a difficult extinction deadline for the species. But the cooperative effort seems to be working. In Turkey, fishermen have witnessed a change in the stability of the coastal ecosystem as well as in their own lives. A gradual improvement in fish stocks has occurred in concert with protection of the seals. And with more fish, both human and seal families can be sustained without conflict.

As if rewarding the groups for their efforts, Poseidon revealed an oceanic secret in January 2011. It rocked the world of monk seals. On an isolated beach, on a lone island off the coast of Greece, a secret colony of Mediterranean monk seals was discovered. Scientists making the unusual discovery refused to reveal the location of the animals for fear of human disturbance. They did, however, share one important finding. They noted that these seals acted differently than any other local monk seal. Instead of hiding in caves, afraid to show themselves to man, as was considered the norm, these secret seals lay about on open beaches, sleeping confidently together as Homer once described the marine mammals during the reign of Proteus. Clearly, monk seals were a different animal when not pressured by human disturbance, and could easily live in the presence of man. More important, if protected, monk seals had the power to improve the coastal ecosystem.

PEDALING BACK TO THE LAB, I began to wonder if there was a Proteus for Hawaiian monk seals. Certainly, the first king of Hawaii, Kamehameha the Great, had recognized the impact of fishing activities on oceanic resources. By regulating temporal and geographic access to designated fishing areas, he established what could be considered the first formal marine protected areas for the island waters in the early 1800s.

More recently, a Hawaiian Proteus has stepped forward, wearing heavy work boots and a yellow hard hat. He arrived quietly at a government-sponsored meeting in Honolulu in February 2011 to discuss the conservation of monk seals in the main Hawaiian Islands. Dressed in work clothes and carrying a duffel bag, the weathered older man sitting next to me looked like a construction worker who had wandered into the proceedings out of curiosity or by mistake. I was astounded when he took the podium and identified himself as Keith Robinson, "cowboy, farmer, fisherman, environmentalist, and co-owner of the island of Niihau."

His family had bought the island from King Kamehameha V for $10,000 in gold; it is now worth over $1 billion. The 69.5-square-mile, low-lying arid island has the shape of a sleeping seal facing northeast, with a single road running down its spine. Despite neighboring cities on Kauai located just seventeen miles away across a wide channel, Niihau is a reclusive paradise inhabited by fewer than 130 individuals. Only relatives and invited guests are allowed, and there are few amenities such as telephones, automobiles, or power lines, earning Niihau the nickname "the Forbidden Isle." But within this landscape Mr. Robinson and his brother have created an environmental and Hawaiian cultural preserve—with the newest residents consisting of endangered Hawaiian monk seals. Fiercely protective of his land and its plant and animal occupants, Mr. Robinson has allowed few visitors to the island. Monk seals seem to flourish under his feudal care, and where seals were once absent due to human hunting, Niihau now boasts one of the largest populations in the main Hawaiian Islands.

"I've turned away lucrative diving and fishing charters to support the seals," Mr. Robinson informed the group. "The seals are smelly and attract sharks. I would not reestablish them on other islands due to the sharks—it would be bad for tourists." Mr. Robinson estimated that there were fifty to two hundred seals at any one time on his beaches. He also warned that they could easily be pushed out if fishermen were allowed into the surrounding waters. He had a proposal for NMFS.

"Niihau could make up for the resistance by the other islands to the presence of monk seals. I propose creating a sanctuary for the seals on Niihau with the people of Niihau employed to count and protect the seals."

With that, Keith Robinson picked up his duffel bag and left the rest of the room in stunned silence. No one spoke except for the woman sitting next to me. "People call him Robinson Crusader," she whispered. "Rumors say he keeps a machete and Glock pistol in that duffel bag."

His quirks notwithstanding, I was fascinated by Keith Robinson's observations about the influx of sharks with the arrival of seals to his island. It seemed to me that there was a much larger ecological story for monk seals and sharks than simply being outcompeted by the faster scavenger. Food webs and species interactions were likely more interconnected than that.

Several scientific studies have already demonstrated the importance of nutrient transfer by highly mobile marine animals. Seabirds are known to deposit guano to islands that act as plant fertilizers where they nest. Similarly, whales and seals carry deep-sea micronutrients to the water surface as they come up to breathe and digest their prey. The feces of marine mammals are natural fertilizers that enable phytoplankton, the microscopic drifting plants that live near the water surface, to grow.

But there is a much more exciting aspect to this nutrient cycle. Phytoplankton provides food for zooplankton, which serves as food for small fish, which in turn become food for middle-sized and big fish.

Phytoplankton, with the help of micronutrients found in marine mammal feces, converts sunlight and carbon dioxide to create food for supporting the entire ocean ecosystem.

Suddenly, I realized that monk seals were not simply fish *consumers*—they could actually serve as fish *farmers*! Their feces would provide critical nutrients to seed coastal waters, and nowhere was this more important than in warm waters where nutrients were quickly lost. Maybe that was the reason that fishing grounds improved in the presence of seals in the Mediterranean, and why sharks were attracted to areas where there were seals here in Hawaii. That was where the fish farms resided.

There was one more piece to this food web that seemed too astonishing if it was true. If the feces of marine mammals fertilized phytoplankton and phytoplankton consumed carbon dioxide in order to grow, then theoretically this nutrient cycle would have an impact on the presence of atmospheric carbon. Such a connection had already been proposed for the great whales. In 2010 scientists reported in the Proceedings of the Royal Society that the feces of Southern Ocean sperm whales were responsible for levels of phytoplankton growth that resulted in the consumption of two hundred thousand tons of carbon dioxide each year. Obviously, the whales produced carbon dioxide with each exhalation. However, the effect of their feces on phytoplankton was much greater. As a result, the sperm whales were responsible for a net loss of atmospheric carbon dioxide that was equivalent to the removal of emissions from almost forty thousand passenger cars every year. Simply put, the mere presence of marine mammals could have an impact on global warming.

Whether science would prove that monk seals serve as fish farmers or a solution for climate change remains to be seen, but the prospect of it sent a shiver of excitement through me. We needed to study this. One thing was already clear: like all large mammals, Hawaiian monk seals were a critical component to the coastal ecosystem. We did not yet have all the answers. But as the only marine mammal in Hawaii to straddle both the oceans and the land, there was little doubt that monk seals had a unique role as the glue that held the islands' coastal ecosystem together.

KP2 and his species needed to be viewed in their true biological role. The seals were neither invaders nor gluttons. Rather, they were ancient, honored family members who helped to create and maintain the tropical coastal ecosystem that both man and seal needed to survive.

REENERGIZED FROM all the thoughts swirling around my brain, I realized that I had been thinking too small about KP2. There was a whole ocean of possible homes for the seal in the islands, an ocean that could use his help.

20.

Wild Waters

In the best of all circumstances, I'd like KP2 released in the Northwestern Hawaiian Islands," I announced to Amy and Jennifer, even though I knew it was a long shot. "Or maybe an ocean pen like he had in Kaneohe. Imagine being able to conduct all types of open-water observations as KP2 hunted for fish. We could determine what kind of fish farmers these monk seals really are. It would be *exciting*!"

To their credit, the two women never shot down one of my ideas. They didn't need to. There was a simple test that would determine if release back into the wild was even worth discussing for KP2. First, he had to pass an eye exam.

The underlying reason for KP2's journey to the mainland and my lab had been for the care of his eyes. He arrived with both corneas veiled gray and cloudy. For nearly two years my team had medicated, shaded, and documented the cloudiness of his eyes. We discovered that intense sunlight aggravated his condition. On foggy days, his brown irises shone bright and clear and his visual perception seemed more acute. In contrast, bright sunny days caused his eyes to appear milky and the seal relied on other sensory cues to help him orient. The question remained: had our efforts and his continued development been enough to stabilize his eyes and stave off surgery, or would the seal eventually go blind? KP2's fate hung precariously on the health of his eyes.

Regardless of the passage of time and all his scientific contributions, possible paths for the seal had changed little since he'd been captured at Kaunakakai Wharf: placement somewhere in Hawaii, captivity on the mainland while awaiting eye surgery, or euthanasia if his eye condition indicated a lifetime of ocular deterioration and misery. In my heart I wanted KP2 to be released in the tropical waters of Hawaii. Even with cloudy eyes he would be able to hear the rustle of palm tree fronds in the trade winds and use his whiskers to hunt for fish. I decided that I would explore every avenue if it meant a chance for him to dive among the corals and again feel the soft, warm sand of an island beach.

Moreover, I wanted the seal to reexperience the luxury of the warm blue waters that had once been his home. Perhaps it was the swimmer in me that made me think this way. As a lifeguard during my college days, I discovered the lightness and power of water, and have always wondered if marine mammals appreciated it, too. Water caresses your skin as you swim, and buoys you physically and emotionally. I imagine the pure joy of being able to perform acrobatic leaps like a dolphin or high-speed, agile maneuvers like a seal or sea lion. Why else would they race through the water in such energetic bursts, if not for the sheer fun of it? I believe that swimming prowess is the secret behind the dolphin's smile.

STANDING IN KP2's humid sealarium with water dripping on me and the ophthalmologist's instruments, I sweated more in anticipation of the results from the eye exam than from the temperature. This was it. We had brought in the best marine mammal ultrasonographer in the world. The next phase of KP2's life hung on the results of Cynthia Kendall's examination.

"He reminds me of a patient with cerebral palsy that I examined last week," Cynthia noted as she tried to position a pen-sized ultrasound probe on KP2's left eyeball. Eschewing any type of sedation, Traci and Beau had worked for months training KP2 to lie quietly for the eye exam. We waited expectantly to hear why the ophthalmic ultrasound specialist likened the young seal to a teenage boy with a neurological disorder. She had tied her auburn hair back and was squinting through thick glasses as she studied the ultrasound screen.

In the few short hours since I had met Cynthia Kendall, I came to appreciate that she was filled with quirky observations. Trained in engineering, physics, and human health and possessed of a fondness for animals, the wiry woman moved quickly between steely objectiveness and patient empathy. She had already revealed some interesting perspectives: the karma of owning a one-eyed feral cat, her love of all things ocular and therefore disdain for Lasik surgery, her self-nomination to president of the KP2 Fan Club, Sacramento Chapter.

"My business card even has a picture of KP2 on it," she said proudly. Noting my confusion when I looked at the black-and-white image on her card, she added, "That's his eyeball!" With the long, thin fingers of an artist, she turned the card sideways and I realized that I was looking at an ultrasound of the cataract she had found when KP2 was just a pup in Hawaii. That single image had prompted KP2's transpacific flight to California and my lab.

Innocence in the face of the ultrasound eye probe was the connection in Cynthia's latest observation. Both the seal and the palsied teenage patient had happily met the ultrasonographer with her $60,000 portable ultrasound computer and eye probe. It was all smiles and curiosity until she proceeded to press the probe onto their respective eyeballs. Shock was replaced by resistance, followed quickly by acceptance, with the realization that the odd sensation of the pulsing ultrasonic probe was painless.

"Hoa, relax," Traci murmured, placing her hands along either side of the seal's head to steady him. With the trainer providing calming words of encouragement, KP2 lay quietly on the deck of his sealarium as the most important eye exam of his life was conducted. No human could have been as composed.

"How big is his eye?" I asked. The bony brow of the monk seal did not have the bulging look of the Weddell seals I had studied in the Antarctic. Those polar seals had enormous eyes.

"I'd say about thirty-five millimeters in diameter." Cynthia confirmed my impression as she rocked the probe over the surface of KP2's cornea. This also added to my suspicion that Hawaiian monk seals might be less adept at diving than Weddell seals were. While the Weddell seal eye was nearly the size of a tennis ball, the monk seal eye was only half that diameter, closer to a Ping-Pong ball. Large eyes contributed to light sensitivity. Anatomically, Weddell seals had two headlights that enabled the diving hunter to locate prey under the extreme low-light conditions deep beneath the Antarctic sea ice. Here they shared the same big-eye feature of other deepwater predators such as the blue whale, with an eye diameter of 150 millimeters, and the giant squid, with the biggest eyes in all the animal kingdom. At 270 millimeters (nearly 11 inches) across, the eye of the colossal squid is bigger than a regulation soccer ball. Monk seal eyes pale by comparison, suggesting adaptation for shallow-water hunting or at least reliance on other, nonvisual sensory cues.

"Oh, ho, there it is. Yes!" Cynthia gleefully pointed out a white marble in the middle of the gray screen.

Images on the computer monitor materialized and dissolved as Cynthia moved the probe across KP2's right eye and then his left. The story was the same for each: a hard white marble where a clear lens should have been. A sinking feeling began to wash over me.

"I think the cloudiness has progressed since he was a pup, just as I'd expect it to over time. But it is stable."

"So how well can he see?" I asked tentatively.

"Even with surgery he'd only be able to make out movement and distant objects. He has little near vision and it will only get worse as he grows older."

"Why? How did this happen?" I felt my dream of releasing KP2 back to the wild melt in the heated sealarium. I demanded to know whether the cooperative seal lying at our feet—a seal who would never again know the freedom of open water—was a genetic anomaly, a victim of an accident as a youngster, or if there was anything we could have done for him.

"My impression? Nutrition!" Cynthia replied without hesitation. "Let's face it. Nothing is better than mother's milk!" According to the ultrasound specialist, for all the calories provided by the salmon shakes that nourished KP2 in his first days, they were not enough to allow his eyes to grow healthy. Scientific literature is filled with studies championing the developmental benefits of breast milk in humans. The signature fatty acids in mother's milk are critical nutrients for proper brain growth and nervous system development. Breast-fed babies perform better on intelligence tests and display superior eye function than bottle-fed babies. Nature knows best, and RK22's legacy of abandonment would forever be imprinted on KP2's eyes in more ways than one.

The seal's original veterinary ophthalmic team, led by Dr. Carmen Colitz, prepared the final report on KP2's eyes using Cynthia's ultrasound images. Carmen had traveled the world performing eye exams and cataract surgeries for sea lions and seals from SeaWorld to the Houston Zoo and the Waikiki Aquarium when KP2 was a pup. She had tracked his case from the very beginning and was kind when she noted my team's disappointment. We had been battling a developmental defect and had just lost the bid for KP2's freedom.

"But you've done an excellent job with his eye care!" Carmen praised my group. "KP2's healthy, and the shaded enclosure you built was perfect for preventing his corneas from clouding over. In fact, his eye health is so stable that we can forgo surgery for the immediate future."

Reluctantly, I accepted the news. We would have to create another life path for KP2.

Two weeks later Cynthia sent me copies of the ultrasound images for my files. After a cursory glance I filed them with the rest of the seal's medical records.

A few days passed before Cynthia called. "Well, did you see it?" she asked excitedly. "Look at the pictures!"

Opening the files, I stared at the black-and-white ultrasound images. All I saw were the depressing white marbles that prevented KP2 from living on his own in the wild.

"Look closer at the picture of his left eye."

I turned the image to the side and then started laughing. Both eyes of the seal showed opaque areas with the thickening of his lens starting on the edges of his eyeball and gradually moving inward. KP2 was essentially seeing through hollow marbles that would fill in as he grew older. But the shape of the hollowed areas differed for each lens. Cynthia had spotted an unusual feature in KP2's left cataract.

"He has a heart in his eye!" Cynthia exclaimed. Indeed he did. The dark area representing the clear part of KP2's left lens was heart shaped and as obvious as any valentine.

As if we didn't already know that KP2 was blessed with a unique heart, he clearly had one imprinted in the hollow of his left eye to remind us.

ALTHOUGH DISAPPOINTED, I realized that I would settle for a healthy seal. Caught up in what I wanted for KP2, I had forgotten that he faced limitations in his vision every day. The seal was so good at masking those limitations through the use of his other senses that we rarely considered him disabled. He operated with an internal map of his enclosure in his head. The map made him an amazingly mobile animal in controlled environments, but would be insufficient for keeping him safe from unpredictable dangers in the wild waters of Hawaii.

One of the biggest threats in the Northwestern Hawaiian Islands, where I had wanted to place KP2, is pollution. Floating marine debris in the form of discarded filament lines, fishing nets, plastics, and ropes creates a lethal obstacle course for swimming marine mammals in the most remote coastal waters on earth. Each year more than fifty-two metric tons of marine debris accumulates on the beaches, on coral reefs, and in shallow lagoons of the Northwestern Hawaiian Islands. Lost and abandoned fishing gear—particularly large trawl nets and fish nets—is the biggest danger.

Hawaii is by no means unique in the level of its marine pollution; it is just surprising (and in many ways disheartening) given the remoteness of the islands. Around the world, the switch from natural fibers such as cotton and linen webbing to more durable plastic fishing gear during the past thirty-five years has created an invisible water hazard. Even sighted marine mammals have difficulty detecting floating nets and become hopelessly entangled. Swimming against razor-sharp filament lines, dolphins have been known to accidentally amputate their dorsal fins and flukes. Ropes looped around the necks and bodies of seals and sea lions eventually cut into the skin and strangle the animals.

"Why can't someone create strong, dissolvable nets and fishing lines?" I lamented to the fishermen in my family. There had to be a better solution than simply allowing lost fishing gear to drift unimpeded across the waves.

"They can. But who would be crazy enough to use it?" they said, laughing at me. "The beauty of plastic is that our fishing gear lasts forever!"

There was no attractiveness in it for me. Such "beauty" came at too hefty an animal price.

Over the decades of my scientific career I had witnessed the impact of marine pollution on the ocean's animals and was driven to solve the problem. The level of devastation from pollution is both astounding and unnecessary. Yet few are aware of the impact, due to the remote location and cryptic behavior of marine mammals. I realized this when directing the rescue of hundreds of oiled sea otters following the 1989 Exxon *Valdez* oil spill in Alaska. The clash of Big Oil, big business, the lack of federal and state preparedness, public outrage, and guilt had left otters dying in my hands. The event had been a turning point in my life.

The drive was reinforced when I moved to the Hawaiian Islands immediately after the spill. Less than a week after my arrival in Kaneohe to work with Puka and Primo, a call came in to the Dolphin Systems veterinary office. A whale had been seen struggling off the coast of Kauai; it appeared to be spouting blood.

By the time the veterinary team and I arrived on the island, the immature sperm whale had stranded and died. The whale's last breaths occurred on the opposite side of the island where KP2 would be born eighteen years later. Its massive heart stopped beating on the prime bathing beach of the prestigious Sheraton Kauai Resort, right in front of the $350-per-night ocean-view rooms of mainland honeymooners.

The resort manager was frantic. The staff at the front desk was visibly shaken, and a depressing air weighed down the tropical fragrance of ginger in the front lobby. A beautiful, statuesque Hawaiian woman working at the reception desk interpreted the whale's death as a bad sign.

"Why did this happen?" she quietly asked, seeking a deeper meaning for the decision the ocean had made in depositing the whale on their beach. "Perhaps the hotel should not have been built," she suggested. "The islands are getting too crowded even for the animals of the surrounding seas. Maybe the oceans are trying to tell us something."

Wringing his hands, the manager asked us what our next move would be. One of the veterinarians mentioned that we needed to necropsy the whale to determine the cause of death.

The manager blanched. "Here? You want to cut the dead whale here? No, no, no!" He then disappeared into a back room, groaning as he left.

It took only a few steps from the lobby to locate the body, resting on its side with its great flukes lapping in the gentle waves. The head of the animal was nearly half its total length. Its stubby pectoral fins stuck out awkwardly from its sides. I was about to place my hand on its gray ridged skin when one of the locals in a flowered shirt stopped me.

"Is someone going to offer a prayer?" he asked.

"A prayer?" I repeated, looking to the rest of the team for help.

"It is customary to say something before you perform a ceremony," a resort staff member clarified. The team thought a moment and gave the responsibility of the prayer to the lead veterinarian, who offered a few words in Hawaiian before our inspection of the body.

I had never seen a sperm whale from so close up. The massive animal's world was the mysterious oceanic trenches that scarred deep into the earth's crust. In total darkness and at crushing depths the whale lying in front of me would have hunted the giant squid. Now one side of the whale's body was shredded; the skin hung loosely where it had been scraped along the black, razor-edged volcanic rocks curbing the beach. I found the dark gray skin surprisingly delicate for such a large mammal.

By the time we walked around the whale once, the resort manager had found a solution for disposing of the enormous carcass. The owner of a nearby sugarcane field would allow us to borrow heavy equipment to move the whale and then bury the remains in one of his fields.

Moving a dead whale is no small feat, and this one was located between the expensive hotel and a lagoon surrounded by crumbling volcanic cliffs. But within twenty-five minutes, a large, steel-reinforced crane and flatbed truck arrived. The crane operator, a muscular Hawaiian with brown chest and belly protruding from a loose shirt, whistled when he saw the situation.

After assessing the problem, the crane operator lowered a cable, which we caught and quickly secured around the whale's peduncle by its flukes. Slowly the crane began to pull. At first the whale did not budge; its dead weight and gravity fought the crane. Eventually the flukes cleared the water, and then the hind end flexed up. Billowing blue smoke from a stack, the crane lifted further as the steel cable strained and groaned under the dead weight. Finally it was able to raise the body and lastly the head of the whale.

The body of the whale swayed slightly in the wind as the operator slowly swung the enormous carcass up and over a cliff edge. In a remarkably gentle move, the Hawaiian operating the crane lowered the carcass of the whale lengthwise along the flatbed truck. It could not have been done with more care than a relative lowering a loved one into a final resting place.

We drove for half an hour up toward the center of the island into the middle of acres of sugarcane. Deep red dust choked us as we followed the truck and dead whale along the rutted dirt road deep into the field. At last, the whale was laid in a clearing of brick red dirt surrounded by green swaying sugarcane. From this point we could see the turquoise waters of the Pacific Ocean on two sides of us. It seemed a fitting prism-colored burial place for a marine animal whose ancestors had once lived on land more than fifty million years ago.

The team spent the rest of the afternoon under the unrelenting sun working on the whale. We checked the whale's skin for telltale signs of parasites, took samples of saliva and blowhole mucus for culturing in the laboratory. Huge flensing knives on wooden handles were brought out to cut through the tissue and bone. We inspected the blubber for signs of starvation and the giant heart for congestive failure. The lungs were the only internal organs to show any signs of abnormality; there was a slight indication of pneumonia.

Sun-darkened cane field workers came to watch and eventually became bored, leaving us to our grisly task. The sun, too, faded off into the west, casting pink and purple shadows over us and the remains of the whale. We were soaked in blood and the fluids of the whale as well as our own sweat. There was little left to do, although we were frustrated and unsatisfied with the progress of our necropsy. Unless the lab reports showed something unusual, there was no obvious reason for this young whale to have died so suddenly. He was healthy. So why, we queried each other, would a healthy whale throw itself on a beach to die?

Only one more organ awaited our inspection: the stomach. We had delayed this unpleasant task to the last moment. The stomach juices of whales are unbelievably foul, and all of us dreaded the oils and acids that would permeate our skin and clothing for weeks in spite of numerous washings and bleachings. Whenever our hands would pass near our face for the next month, we would detect the telltale smell of bile and whale carcass. Our plan was to evaluate the stomach as quickly and simply as possible.

Light was fading fast on the tropical island and the headlights of several trucks were turned on to illuminate the exposed side of the whale. They cast eerie shadows on the carcass, and the words of the woman at the reception desk of the resort haunted me in the cane field. Maybe the oceans *were* trying to tell us something.

We decided to avoid a big, oily gastric mess by making a three-inch-incision along the side of the whale through the blubber and muscles overlying the stomach and finally into the stomach itself. Because I had the longest and thinnest arms, I was elected to feel around inside the stomach and report what I felt back to the group. (In truth, I believe the veterinary team saw this as my initiation rite to the islands.)

I swallowed and resolutely stuck my hand and arm into the slit under the glare of truck headlights and with the wooden sound of cane stalks rustling all around us. At first there was nothing much to feel, smooth tissue edges and slippery fluids. Then something hard and slithery passed by my fingers; I couldn't tell what. In the dark and without being able to see my arm up to the elbow, I first thought of parasites. Nematodes, cestodes, and innumerable, indefinable spineless worms that I despised raced through my mind.

Squeezing my eyes shut, I tried to concentrate on what I was feeling. Suddenly, my hand reached something that had more substance, something hard that didn't slip loosely out of my grasp when I squeezed down. I decided to bring up what I could and quickly dump it into the bucket of water at my feet. In a single motion I grabbed and threw everything within my reach into the bucket.

In the dark shadows the team couldn't tell what had been brought up from the whale's stomach. At first it looked like the partially digested tentacles of an octopus, and then some type of elongate brown worms. The veterinarian who had performed the blessing took a piece of dried cane stalk and began to probe the brown ball, trying to unwind it. The mystery unraveled with his probings. What I had grabbed was not biological at all; it was man-made. The long tentacles turned out to be rope. I had extracted rope and nylon-filament twine from the stomach of the young whale. There were yards and yards of fishermen's netting. It was the kind of netting thrown into the oceans to drift aimlessly on currents to catch squid until the owner retrieved it at a later date.

Wide-eyed, we quickly pieced together the demise of the young sperm whale lying at our feet in the Hawaiian cane field. Ten pieces of netting were eventually retrieved from the animal's stomach. Thick brown twine nets, razor-sharp filament-line nets, as well as large nets and small pieces of net had tangled together. In searching for food, he had learned that the easiest squid were found entangled in these drift nets. But the filament nets were invisible to his underwater X-ray type sonar detection system. He had feasted only to have the indigestible nets ball up tightly in his stomach. Even the acids of his digestive tract had been unable to dissolve the plastics. Eventually his inability to see man's handiwork had cost him his life.

THE SAME NETS AND PLASTICS landing daily on Hawaii's beaches became beds for monk seals that eventually strangled necks and limbs in boa constrictor fashion. Monk seals, with their innate curiosity and peculiar habit of cuddling beach trash, were especially vulnerable. KP2's species had the unfortunate distinction of maintaining one of the highest documented rates of entanglement of any seal or sea lion species on earth. Encounters with marine debris were routinely reported across the islands, and the population was hurting as a result. The same month that I met KP2 at the Waikiki Aquarium, a male monk seal pup was born at Koki Beach on East Maui. During his first swimming lesson with his mother in the shallow coastal waters, the pup encountered a ziplock plastic bag. It was the kind typically used by snorkeling tourists to hold bread to feed the reef fish. Out of curiosity, the pup nuzzled the bag only to have the plastic top ring slip around his neck.

Although he was not in immediate danger, if the bag had remained, the ring would have tightened and eventually cut into his skin as the pup grew older. I had seen this happen with all types of plastics and filament lines on marine mammals around the world, from California sea lions near Santa Cruz to sea otters and Steller sea lions in Alaska. The same was true for fur seals in Africa, and for marine mammals from the Arctic to Antarctica. Plastics were ubiquitous in the oceans.

As the plastics slowly slice into skin, the victim dies from either infection or from sharks, attracted to the scent of death, that attack the weakened, bleeding victim. The monk seal pup from Maui had been lucky. He eventually figured out a way to swim backward and pulled out of the plastic ring.

With his diminishing eyesight, KP2 would not stand a chance against plastics floating along the coastline. My dream of release for the seal had become hopelessly entangled in the marine debris issue.

Yet all was not lost. According to the elders of Molokai, destiny had another plan for the young seal. He was Hoʻailona, a sign. Although he was unable to return to the oceans, KP2 could do something no other monk seal could accomplish. He could make the waters of Hawaii a safer place for the rest of his wild family.

"Okay, Hoʻailona," I informed KP2 after his eye exam. "It is time for you to live up to your name and be a voice for the oceans."

21.

The Voice
of the Ocean

⟿

KP2 as Hoʻailona immediately instituted a war against plastics and marine pollution on his Web site and Facebook page. He asked young students to clean up their schools and adults to clean up rivers and beaches. "All water is connected," he reminded them over the Internet.

One of his first targets was plastic bottles. Recognizing the landfill overburden created by disposable water bottles and juice containers used in schools, the seal called for a reduction in disposable plastic containers by children. He offered to send a sticker with his picture on it to anyone switching to a reusable bottle.

To illustrate his cause, I attempted to conduct a photo session with KP2. My idea was to have the seal pose next to my Nalgene bottle decorated with one of his stickers. What began as a photo shoot ended as a soccer match between me and the seal. To my surprise, KP2 found the Nalgene bottle irresistible. After posing politely next to the bottle, he proceeded to drag it into his pool as if it were a newly discovered coconut.

"Oh, no, you don't!" I countered, and tried to scoot the bottle back onto the deck. The seal anticipated my move and threw a body block on top. Maneuvering around the immovable seal, I tried to find his weak spot. He couldn't roll the bottle without opening his guard. "It's just a matter of time, old boy," I challenged him.

But the crafty seal was faster than I imagined. Positioning his chin on the bottle as if he were giving up, KP2 hugged the bottle and rolled his body toward the pool. If he reached the water I'd never stand a chance of retrieving the bottle, so I lunged for it. The seal's only miscalculation was the slipperiness of the plastic. He squeezed a little too hard with his front flippers in his getaway and the bottle shot back within my reach.

Unlike the stepladder tug-of-war, the scientist won this time. But the battle also emphasized the link between monk seal behavior and marine pollution. KP2's game demonstrated the exceptional curiosity of monk seals and how quickly they were attracted to novel garbage in their environment. In this regard, they were like the Alaskan sea otters that were so curious about new things in their habitat that they were actually attracted to the crude oil during the Exxon *Valdez* spill. Monk seals, including KP2, had a similar fatal behavioral flaw.

LOCAL SCHOOL GROUPS soon picked up KP2's cause, especially in light of the recent discovery of the Great Pacific Garbage Patch, a gyre of discarded plastics and debris that swirls for miles in the Pacific

Ocean. One class stood out in its unwavering dedication to the problem. Jessica Cambell's fifth-grade class at Mount Madonna School in Watsonville, California, not only *wanted* to stop marine pollution, they *were* going to stop it. She e-mailed me with a simple request: "Will you serve as a science mentor for my fifth-grade class project? They have decided to save sea otters from marine pollution."

I stared at Jessica's message in all of its naïveté, considering whether to delete it or pass it off to one of my graduate students. I had adult-sized, global problems to worry about. I had massive bills to pay, permits to write, classes to teach, and animals that constantly needed feeding. Somehow I had to find a grant or the animals and I were going to end up on the street.

With a single stroke of the delete button, Jessica, fellow teacher Sri McCaughan, and the fifth graders of Mount Madonna School would dissolve into the computer ether. My hand hovered over the keyboard. But I couldn't erase the e-mail. I saw my young self with a pocketful of baby birds and mice facing the imposing black penguinlike silhouettes of the nuns shaking their habited heads. The years had switched our roles, and I found the prospect alarming.

A week later I met with the teachers. They had been studying sea otters and decided to organize a beach cleanup near the Santa Cruz boardwalk.

"The kids love the ocean, although most of them do not live near the coast," Jessica said excitedly. "They decided on a project to save the endangered sea otters in Monterey Bay. They've also been studying marine pollution and how it threatens the animals' survival. So they want to clean it up!"

"Saving sea otters is good . . . ," I admitted. "But would you consider a bigger project for your kids?" I launched into KP2's story and explained how entanglement in marine debris was a factor contributing to the demise of his species in Hawaii. I told Jessica and Sri about the children of Molokai and their shared passion for the ocean. "Your students' story is exactly the same as theirs," I concluded. "They also love an endangered species—you have the California sea otter, they have the Hawaiian monk seal."

It didn't take long to craft the framework of a transpacific science project with the school teachers. Before we parted, I added one more piece. Recalling the many sunsets watched from my office, I related my feeling that only the waters of the Pacific separated KP2 from his wild cohort, and that theoretically a water ripple could travel between our coast and the islands. "Maybe the kids in California and Hawaii could clean up beach trash and then touch the Pacific Ocean at the same time," I suggested tentatively to the teachers. "It would be a tangible way for them to connect with one another."

Hence was born Worldwide Waste Reduction Day and the California-Hawaii toe touch.

Jessica and Sri contacted elementary schools in Hawaii and quickly found a compatriot in Diane Abraham, a schoolteacher from Molokai. As an avid paddler, she had experienced a few close encounters with KP2 during the seal's time at Kaunakakai Wharf.

"I had my own up close and personal with him myself," she wrote. "I paddle an ocean canoe, and he thought the canoe was a toy just for him, and put his head on top of it, trying to get on board. I was afraid he would make me *huli* (tip over), so I tried to get him off with my paddle. Wrong move. He thought it was a game I was playing. This went on for a good ten minutes. A friend on shore had to lure him away so I could make my escape. He is a playful guy!"

With the project growing, the response from NMFS in Hawaii became cautionary. "Don't bring KP2 into this," David Schofield warned as he fretted over renewed protests on Molokai against his agency.

With the new monk seal translocation scheme on the table, this was not the time to get people from the main Hawaiian Islands agitated. His outreach assistant added, "You can't have mainlanders telling Hawaiians to clean up their beaches."

"It's too late," I replied matter-of-factly. "It's already happening."

At this point there was no stopping the schoolchildren from talking to one another. Schools from three different Hawaiian islands had already signed up for the beach cleanup and simultaneous toe touch with Mount Madonna School. Barbara Jean Kahawaii's class from Laie Elementary School would be working on the north shore of Oahu. Laurie Madani's fifth-grade class from Kaunakakai Elementary School on Molokai would be touching the ocean at Kaunakakai Wharf. The Kona Pacific Charter School would clean beaches on the Big Island, and Diane and her volunteers would be cleaning another Molokai beach. At the same time, nearly twenty-four hundred miles away, Mount Madonna School would be working on the beaches of Santa Cruz near my lab and KP2.

Sharing conservation concerns across the ocean was eye-opening for both the children and the teachers. Life on the islands added a whole new dimension to the problem of waste reduction that few mainlanders ever considered. Diane sent several e-mails explaining how critical recycling was when you lived in such small quarters as an island in the middle of the Pacific Ocean. She taught all of us that most islanders on Molokai grew their own produce. Her own garden, squeezed into a tiny yard, included spinach, lettuce, radishes, beans, squash, cucumbers, strawberries, *lilikoi* (passion fruit), mountain apple, chili peppers, basil, banana, and papaya. By sharing with neighbors they increased the size of their collective gardening, and on any given day Diane would come home to find a cabbage or some kale waiting on her lanai.

. . .

DURING THE EARLY MORNING HOURS of Worldwide Waste Reduction Day, a marine fog rolled onto Main Beach in Santa Cruz and threatened to dampen the cleanup. When I drove up at nine a.m., the sun had forced its way through the mist and children were busy sifting through every foot of beach sand.

Save Our Shores, a nonprofit group dedicated to protecting the ecology of the Monterey Bay, had outfitted dozens of children with large green pails, trash grabber poles, and plastic gloves. Schoolchildren of all ages scoured the high-tide debris with the intensity of an Easter egg hunt. Plastic bottles, cigarettes, candy wrappers, and much more filled the buckets as the ninth-grade class of Mount Madonna School joined with preschoolers, kindergartners, fifth graders, and their parents.

While the others picked up trash, one of the older girls broke from the swarm of beach cleaners and walked down to the surf zone. She drew a giant heart in the sand and wrote "peace" on the inside. Then the teachers, children, and parents all gathered around the heart holding hands. Young and old were silent as Jessica spoke about the oceans and how humans had an important role in preserving the environment. She asked for their commitment by having all of them place a foot inside the giant heart. After a moment of reflection she instructed, "Okay, line up!"

The children ran laughing to the tide line facing to the west. Thousands of miles away across the Pacific on the island of Molokai, on the beach at Kaunakakai Wharf where KP2 once spent his first birthday in the arms of the Hawaiian children, his old friends also lined up. At eleven thirty a.m. Pacific time and nine thirty a.m. Hawaii time, children of the mainland and the islands ran into the water. Ignoring the chill of the California water, the children splashed and jumped while Jessica spoke on her cell phone to the teachers in Hawaii. In the background she could hear the same sound of children laughing across the ocean.

All of a sudden the mood changed. "Look—look out!" shouted a tiny girl, pointing into the Santa Cruz waves. In front of the splashing children a domed head appeared offshore. Just as quickly it disappeared. Then several minutes later it popped up again a little closer to the group. Parents lined up with their sons and daughters to identify the curious onlooker.

"It's a seal!" one of the fathers identified correctly. "Did it swim all the way from Hawaii?"

The unusual visitor was a cousin to the Hawaiian monk seal that lived down the coast. I'd never seen a harbor seal at Main Beach in Santa Cruz before. The black-and-white spotted seals usually spent their time lolling around the rocky shore and kelp beds of Monterey to the south. So I was pleased that the animal had chosen that moment to make an appearance. His audience on the beach was made up of children who would one day grow up to be the adults responsible for caring for his ocean home. The teachers standing in the sand had ensured his protection by giving their students one special gift: environmental connectedness. The young harbor seal and I had been lucky enough to witness their entry into the fold.

WORLDWIDE WASTE REDUCTION DAY and KP2 spread the word about plastics and marine pollution across the country and across the continents. While the schoolchildren of California and Hawaii splashed on their respective sides of the Pacific Ocean, the Santa Cruz Public Defender's Office and members of KQED in San Francisco picked up street trash that could have fallen into drainages to the coast. Adults and children in South Carolina, Minnesota, Nebraska, Oregon, and Washington State cleaned rivers, beaches, and fields.

In Texas, high school students in Fort Worth were reeling from the aftermath of the Deepwater Horizon spill. Drawn to the plight of marine animals facing man's waste, they sponsored a campus cleanup in honor of KP2. Using the trash they had collected during the cleanup, they separated the plastic bottles, cans, and plastic bags. Taping them together and covering the trash sculpture with papier-mâché, they formed the first life-sized Hawaiian monk seal to visit a Texas high school. They painted their seal with KP2's silvery gray and white coloration and displayed their creation in the lobby of their school to inspire others to reduce pollution. The students named their seal Nani, which they told me meant "beautiful" in Hawaiian.

The inspiration of KP2's message did not stop there. Pictures from conservation-minded individuals poured in from around the world. Men, women, and children sent pictures of themselves and their collected trash from India, Canada, Belgium, Denmark, Australia, Spain, Japan, and Kosovo. They did this for the oceans. They did this without the benefit of ever having met a small seal named KP2.

Afterward, the fifth graders from Mount Madonna School presented their project to the entire school assembly. Among their presentations was an original hula dance for KP2 set to the music of Lono's song "Hoʻailona." Together the children of California raised their voices for their new friends in Hawaii, "Please, oh please, bring Hoʻailona home!"

WALTER RITTE AND HIS WIFE were moved enough to board several planes from Molokai to Oahu to California to visit KP2 in Santa

Cruz. They wanted to see the seal for themselves and learn what we had been doing with him. Known as Uncle Walter and Aunty Loretta at home, the Rittes rarely left the islands and were somewhat daunted by the bustle of San Francisco International Airport and the Bay Area traffic.

I was also anxious about the meeting, unsure what to expect from Walter. To me he was the "Molokai activist." I was afraid he would see KP2 in our pools and contact the media about scientists imprisoning the monk seal that once swam freely in the waters of Molokai.

Instead, what the Rittes saw was KP2 teaching children. When they walked into the marine mammal compound, KP2 was sitting on the walkway with Beau in front of a line of spellbound schoolchildren. Each child had a notebook and was busily writing down data as Beau measured the seal's skin temperature along his body. They were studying the monk seal's "homeless pose" and determining if temperature regulation was related to the species's odd habit of snuggling beach trash.

Walter and Loretta were incredulous; they had never seen anything quite like it. In Hawaii, KP2 had been a playmate and a comic. When other wild monk seals came up on public beaches, NMFS volunteers would often cordon them off with yellow police tape to keep people at a distance. Even in aquariums, Hawaiian monk seals were behind glass or across moats to prevent human contact. Here was a seal where there were no barriers. The Rittes immediately recognized the empathy for the seal in the children's eyes.

Once the lesson was over, the children thanked KP2 as he followed Beau back to his pool. Alone with the seal at last, Walter and Loretta reacted with the heartfelt cheer of any reunion with beloved family.

"Oh, KP2, there you are!" Loretta crooned as she raced up to the

Plexiglas partition of the sealarium. Walter stood behind smiling widely.

Hearing the visitors, KP2 surfed across his pool and popped onto the deck. Without hesitation he shoved his muzzle on the partition in his usual whiskered greeting. Walter watched without comment with the seal clearly focused on his movements.

I stood off to the side and could see KP2's curiosity with the island visitors. He always greeted folks, and then after a few moments went back to swimming. This time he stayed. He listened to the sound of their voices and sniffed the air through a small gap in the bottom of the wall separating him from them.

"Maybe he can smell the island of Molokai on you," I suggested with a laugh.

Walter considered my comment for a moment and then bent down to slowly exhale into the same gap in the wall. For several minutes the seal and the islander breathed each other's air.

THE VISIT BY THE RITTES served to break down so many barriers and misunderstandings that had prevented communication between my lab and the islanders. Walter and Loretta taught me about Hawaiian culture and the relationship between the islanders and the oceans. In turn, we taught them about the science of saving an endangered species. The transition was easy once we recognized that we were connected by a common love of Hawaiian monk seals.

We also shared a common insight about the special orphaned seal in front of us. KP2's ocean message and his new role as teacher and ambassador were key to the survival of his species.

22.

Aloha Hoa

———

Months later, I received a phone call.

"We need KP2." Walter Ritte was whispering so quietly on the phone I was unsure I had heard him correctly. "He has been blessed as Ho'ailona, and he needs to speak for the oceans." The man sounded tired.

Walter had been talking with Karen Holt, the executive director of the Molokai Community Service Council, when they decided to call me. For months NOAA had been collecting opinions and getting an earful from local Hawaiian fishermen across the islands voicing their concerns about monk seals and the government agency's proposed translocation of seals to the main Hawaiian Islands. The idea of flying newly weaned female pups hundreds of miles south, from areas of low juvenile survival in the Northwestern Hawaiian Islands to areas of high survival in the main islands, was unprecedented for the species. The fishermen were skeptical about the ability of NMFS scientists to recapture the grown seals several years later and return them to their original capture sites.

Things were going badly for the seals.

"People do not trust NOAA, the government," Karen informed me on the conference call with Walter. "They say NOAA has allowed too many fish to be taken from our waters. The government brought the mongoose to the islands and let them take over our fields. They allowed giant fishery corporations to move into our waters. No one has been honest with us." She sighed in frustration. "We don't know who to believe anymore."

"No one is speaking up for the seals," Walter added. Indeed even state politicians were echoing the sentiments of their fishermen constituents.

According to Karen and Walter, those favoring the translocation were noticeably absent from the town hall meetings. NOAA's radical idea of shuttling seals around the islands had provoked images of scientists playing God. Local sentiments were clear: fishing grounds were too sacred for such interference.

There was one group, however, whose small voice struggled to be heard. Although nearly two years had passed, the children of Molokai never forgot the friendly seal pup who once swam with them. They had grown up learning about the importance of conservation and how each of them left a footprint on the earth. Such concerns had inspired beach cleanups and a ban on plastic bags and disposable water bottles. The children were writing letters of support for the seals to anyone who would listen.

"We need KP2 to come home now," Walter finally urged. This time there was no demand to bring the seal to live in a fishpond on

Molokai. The issue of conserving Hawaii's designated state mammal was bigger than that. KP2 needed to go where his public impact would be the greatest.

In the past, I would have repeated that there was no place for KP2 to live in Hawaii. I had gradually come to terms with and accepted that his eye condition prevented his release back into the wild. I knew that a special facility was needed to care for him. This time, however, I had good news for the people of Molokai. We had found a home for KP2. In another twist of fate, Nuka'au, the seal that once shared a pool next to KP2 at the Waikiki Aquarium, had passed away. He had lived to be thirty-one years old, the equivalent of ninety years for a human. With his peaceful departure, KP2 unexpectedly had a Hawaiian home waiting for him on Oahu.

"Your seal is finally returning," I happily replied.

FROM AN UNCERTAIN FUTURE, we suddenly had only one month left with KP2. My team began a mad scramble to finish everything we had planned to do with him in addition to our preparation for the seal's big return to the islands. After all our struggles, I could not believe that our time with the endangered seal was nearly over.

It had been a long two years in which we helped KP2 transition from wild pup to maturing adult. As I added up the milestones and the many scientific accomplishments, I knew that all could be attributed to one key element: *hilina'i* (trust). Starting with Beau, my team had developed complete trust in KP2, and in turn KP2 completely trusted us. This once wild seal had learned to sit still for weighing after only one day of training and had overcome his fear of needles in less than two weeks under our care. We had learned not to put any behavioral obstacles in his way. KP2 had demonstrated his remarkable intelligence and along the way taught us about the biology of his growth, metabolism, thermal preferences, and the energetic cost of catching a fish. No other monk seal had contributed so much for his species.

All of KP2's data were destined to enter ecological models of the Hawaiian environment to predict the habitat needs of his wild cohort. Marine reserves could now be designed based not on what humans outlined on a map but on the unique biology of this warm-water seal. Managers would now know what water temperatures and what water depths were best for the seals' survival. They could accurately predict how many fish were needed each month to support the seasonal fluctuations in growth, molting, and reproduction of monk seals. Because no such data existed for his sister species, KP2's research would also provide insight into the habitat requirements and biological limitations of the highly endangered Mediterranean monk seal. With help from my team, KP2 had more than proven that it was possible to aid wild populations through research on an individual under human care, on an animal once considered an expendable castoff to be euthanized.

There were so many landmark achievements, but I still wanted more before KP2 left. Recognizing that I might never have another chance like this again, I asked for the unspeakable. I wanted one last data point from KP2.

Over the years I had learned never to verbalize this wish to Beau and Traci. There is a superstition among research trainers: never say that you are collecting your *last* data point. If you do, something inevitably goes wrong. I should have heeded that advice.

As I rode my bike along the coast to the lab, I considered the final data session I wanted to conduct with KP2. Traci and Beau were waiting with him so that we could discuss research priorities; there simply were not enough days left to do everything that I'd planned. I had to smile. That was always the case in science. Answering one question always sparked new queries.

I pedaled faster up a grade and then glided down the decline like a diving seal. Traffic was light and I waved at a passing cyclist on the other side of the road who had also taken advantage of the spectacular autumn sunshine. The day was perfect.

But before the other cyclist disappeared behind me, a black BMW suddenly swerved across the traffic and came to a halt in my direct path. "Why doesn't he *move*?!" was the last clear thought I had that day. Reflexes and thousands of miles on bikes presented two disagreeable choices that had to be decided in a microsecond: hit the car broadside or brake for the best. I chose the latter.

The desperate move propelled me over my handlebars. I felt the hard crunch of my helmet on asphalt. Just from the sound, I knew that I was in the worst accident of my life. Instinctively, I did not move. There would be no getting up from this one.

Faces came and went. I remember a young kid with a cell phone—the driver of the car? A middle-aged man with brown hair seemed to take charge. A gray-haired woman hovered nearby. Sometime later—one minute, ten minutes, a half hour, I couldn't tell—a heavyset man in a khaki uniform asked questions. I wonder if there were sirens. At the time it seemed to me that sirens were usually associated with scenes like this. I never heard any before I blacked out.

I SUPPOSE IT IS INEVITABLE when you are strapped to a backboard in an ambulance being rushed to the hospital that you review the events of your life.

All I could think of as I faded in and out of consciousness was, Is this how it is all going to end? Am I just going to blink one last time and everything that I've tried to do will just vanish?

Staring at the ambulance roof, I felt no fear. I let the people around me do what they needed to do and just considered my life. I was pleased that I had no regrets, except one. What would happen to the animals? It was the one unfinished part of my story.

My mind drifted to the ocean. More than anything I wanted KP2 to have a happy home and a quality life that lived up to his message. I wanted Hawaiian monk seals to survive long into the future and their oceans to always be clear and blue. Most of all I hoped that humans would learn to respect and cherish all animals as partners on this earth. My wish, as EMTs wheeled my broken body into the ER, wasn't to remember my world as it once was—it was a wish for a brighter future.

Several hours and innumerable X-rays and CAT scans passed before my amnesia subsided. My memory was finally jolted by a gentle pat on the hand by a Dominican nun making the rounds through the Catholic-run hospital. "Ach, look at you. *Tsk, tsk,*" she fretted in an Irish brogue. "And no way to call for help."

I was still strapped down with my arms at my sides. The kindly nun placed a nurse station call button into my immobile hand and abruptly left. I hadn't seen a Dominican nun since my childhood. Had it not been for the call button, I would have sworn she'd been an apparition. I wondered if she meant "call for help" in the literal or spiritual sense. From her perspective, I'm sure she thought I was in need of both.

Weeks went by while my bones healed and my head cleared. Over several months orthopedists and physical therapists would make the pieces work together once more. A fractured radius and index finger from my left arm's attempt to stop my fall ripped up one side of my body. On the other side, where I landed on my right shoulder and head, a torn rotator cuff and a concussion kept me aching and dizzy. Muscle spasms throughout my torso made breathing difficult. As the days ticked by, I finally gave in to the injuries and admitted that the door had closed on our monk seal research. Our work with KP2 was done; the last data point would never be taken.

FROM SANTA CRUZ TO HAWAII, the view outside the airplane window never changes. For five hours in a commercial jet or eight hours in a U.S. Coast Guard C-130 lumbering across the Pacific there is little more to see than clouds and open blue-gray water. And just as you might begin to lose faith in the pilot's ability to find the tiny mid-oceanic target, the Hawaiian archipelago appears through the tropical island haze.

It is a heart-pounding, breathtaking view. My heart raced at once again seeing the four distinct volcanic peaks of Maui, the Big island, Lanai, and Molokai. As the pilots banked the plane in our approach to Oahu, we skirted along the northern dorsal fin shore of the dolphin-shaped Molokai. KP2's old Hawaiian home filled the landscape, giving me a sudden appreciation of the oceanic journey he undertook when he was less than one year old.

As we flew below the clouds, the island of Molokai stretched the height and breadth of my horizon. The enormity of KP2's journey at such a young age could only be appreciated at this altitude. I wished he could have had a window seat. Instead he was sleeping in an animal crate tethered with cargo straps to the metal floor of the C-130.

Beau, Traci, and Dr. Casper fussed over the seal in preparation for landing. With my cycling injuries still healing, I decided to let the others do the heavy work. I had already said my "Aloha" to KP2.

THE WEEK BEFORE, I had quietly entered KP2's sealarium after the others had left for the day. In the absence of people, the lab was wonderfully peaceful. Junior and the other cockatoos were quietly tucked in sleep on their perches. The sea otters floated quietly with their mittened front paws folded in skyward prayer. Puka and Primo had drifted to the bottom of their pool to rest with full bellies.

In the sealarium, KP2 was not quite yet ready to sleep and was busily head butting his toys around the pool. He stopped when I walked in, then slowly swam over to inspect the noise. I wondered if he could sense the trauma that my body had gone through. I had little doubt that the dolphins could do so; if they could see through aluminum bars surely they could detect broken bones in humans. Seals were not known to possess such highly developed sonar capabilities. Regardless, KP2 curbed his enthusiastic greeting that late afternoon.

Instead of flopping onto the deck, he sank down in the water and brought up a rubber dog toy on his nose, depositing it at my feet.

"You never were afraid to share, were you?" I asked him, laughing and shaking my head. "You teach them, Ho'ailona," I told the seal as he floated nearby. "Teach them all to share the oceans. Teach them about *hilina'i*—trust—and to have the courage to trust one another. That is the only way we will all survive." Like Beau, I had no delusions about what the seal understood. But as always he seemed to enjoy the company and the sound of a human voice.

I took the liberty of petting the big seal from head to toe as he sidled up lengthwise to the side of the pool. It would likely be my last opportunity to do so. KP2 was one of only two animals in my lifetime who knew how to read me. I was going to miss the animal connection and the marine mustiness of his silvery back as much as I still missed the softness and the smell of Austin's sun-warmed fur. But this time there would be no tears. Scientists don't cry.

As I stroked his warm, slick back, I noticed that there was a lot more of KP2 to love since his arrival two years earlier. It seemed amazing that this two-hundred-pound seal had refused to eat when he'd first arrived at my lab. Every inch of him was handsome. He would always be easy to recognize even in a herd of monk seals. KP2 still had the longest whiskers of any seal I had ever met, and never lost the white angel's kiss patch on his left hip.

After he rolled over I gave the seal a good belly scratch. It was a proper ending to our time together, with KP2 closing his big, dark eyes in doglike contentment.

"One more thing," I warned the seal as I stood up to leave. "Never, ever let the others know that I used to talk to you!"

THE CELEBRATION FOR KP2 began even before the C-130 skidded to a halt on the tarmac at the U.S. Coast Guard air station on Oahu. A crowd of greeters had gathered from across the islands to meet the plane and its precious cargo. They were members of KP2's 'ohana from all walks of his incredible life. On the tarmac David Schofield waited with representatives from NMFS, the U.S. Coast Guard, the Waikiki Aquarium, and the Hawaiian Monk Seal Recovery Team. In honor of his journey and his long-awaited return, KP2 would receive a welcome blessing from the elders of Molokai when he was placed on exhibit at the aquarium. For now, a flower lei was draped on his transport cage as a local priest announced with a Hawaiian chant that their island son had finally returned safely to his family.

As a police-escorted caravan whisked him away to his new aquarium home, KP2 took his first breath of Hawaiian air in two years. Likewise, Beau, Traci, and I inhaled deeply with relief, taking in the fragrant tropical air. Our journey was over. We had attempted the absurd and could rest easy—at least for the moment—in the satisfaction of having achieved the impossible for this one abandoned, endangered seal. Now it was all up to the people of Hawaii to embrace Ho‘ailona's message.

KULEANA IS A HAWAIIAN WORD that has no direct translation into English. It describes the sense of ancestral-based responsibility that often comes with a unique undertaking or experience. It is destiny with a DNA underpinning coupled with a realization that you are doing what you were meant to do in this life, the harmonization of talent and trajectory. I believe that KP2 as well as Beau, Traci, and I were blessed with *kuleana*. It began with the gift of being able to read others—for *kuleana* is best achieved not by looking inward so much as being able to look outward—it grew with the confidence of traveling a clear path.

In my experience, the happiest individuals are those who have discovered their *kuleana*. Such individuals weather hardships, challenges, and sacrifices not as obstacles or excuses for failure but as a natural part of life's adventure. The entire odyssey called life is a joy.

For all the difficulties, Beau, Traci, and I love our work and our lives. We recognize the privilege of being in the company of animals every day. In this regard I could never hate the local fishermen. In their own way they loved the oceans and were following their passion, just as we were.

There was a difference, however. *Kuleana* demands responsibility and respect for your environment. Survival of the fittest, then, could never be interpreted as a license for the most intelligent species to destroy another.

ABUSED AND ABANDONED, harassed and shuttled between homes, slotted for euthanasia on several occasions, KP2 never lost his enthusiasm for life, his fascination with humans, or ultimately his *kuleana*. I found a lesson in that. If more people were able to read the world around them, they would instinctively know happiness. The solutions for the preservation of the oceans, and the conservation of monk seals and the remaining animals of the world, would come naturally.

KP2's ODYSSEY TAUGHT me that when embarking on any path we erase footprints behind us in order to move forward. The days of diving for fish and finding a mate in the wild waters of the Hawaiian Islands were now relegated to KP2's sighted past. Similarly, in trying to save the world's animals, I had lost the link to my own biological future. KP2's and my lineages were destined to go extinct.

Yet both the seal and I were privileged to be living in the present. We had each managed to cross the boundaries that separated humans from animals, and so helped to edge each other's species away from the extinction precipice. That was enough for any one life.

For some, the birth of ideas and the nurturing of inspiration serve as equally important paths to immortality. That was KP2's and my shared destiny, our shared *kuleana*.

Epilogue

In the time it took for us to care for KP2 and for me to write this book, the Hawaiian monk seal population declined another notch. When KP2 arrived in California, his wild family hovered at just 1,100 members. By the time he returned to Hawaii, the numbers had declined to 1,060 seals, maintaining a trajectory for extinction within fifty years.

There have been new threats and new advancements for monk seals as scientists, federal agents, fishermen, conservation groups, and the public wrestle with the next urgent steps. My theory of traveling water ripples between Hawaii and distant land masses became a stark reality with the 2011 earthquake-generated tsunami that swelled across the Pacific Ocean from Japan, washing over the Northwestern

Hawaiian Islands and destroying parts of the Santa Cruz Harbor. Shockingly, a wave of debris including lumber, appliances, and plastics from the Japanese earthquake is scheduled to hit the Northwestern Hawaiian Islands and the monk seals' prime habitat beginning in March 2012. Scientists are preparing for the cleanup.

Traci, Beau, and I are awaiting the arrival of new monk seals at the lab to fill our heated pools and our days with their antics and biology. In a move reminiscent of Morgan the sea otter, we've agreed to try to rehabilitate one of the wild pup-killing monk seal males that would otherwise have to be euthanized to save the population. We still don't have enough money, but with KP2-like determination, we continue to turn over every funding rock to support our research ideas. In the meantime, we are exploring the science of "farmer" seals and staying one step ahead of the university accountants.

THE FEDERAL AGENCY charged with managing the Hawaiian monk seal continues to struggle. As of this writing, the hopes of NMFS scientists to translocate young monk seals to the main Hawaiian Islands were dashed by public and political outcry, and a slashed research budget. The idea has been shelved for the immediate future, although Keith Robinson's proposal to place the monk seals in Niihau is still being discussed. Both the NMFS and the Hawaiian community are looking for creative solutions for restoring trust in one another.

As we had hoped, KP2 has become an integral part of that creative solution. In the same month that NMFS put their translocation plans on hold, KP2 passed his quarantine exam and was finally placed on exhibit at the Waikiki Aquarium. Unlike his poolmate Maka'onaona, who ignores shouting visitors, KP2 has embraced his ambassadorial role. He loves the large glass-fronted pool that offers a window on the human world. On any day he can be seen sunbathing, surfing along his pool, and smashing his giant whiskered muzzle into any hand pressed on the opposite side of the glass, to the delight of both adults and children. Despite my early concerns about his class-clown character, he has matured into the ideal boisterous spokesseal for the oceans and for a species in trouble.

There has also been a glimmer of hope for KP2's wild family.

On April 25, 2011, one week before KP2's third birthday, RK22 again returned to North Larsen's Beach in Kauai. Heavy in pregnancy, she dragged her body onto the warm sands that were once the site of KP2's tumultuous beginnings. His mother gave birth to his newest sibling in the same manner as she had with him and his sister— slippery, wet, and sliding unceremoniously from between her back flippers.

This time, she did not attack or swim off. For the first time in her life, she stayed.

As observers for the Kauai Monk Seal Team once again looked on, RK22 became a vigilant and courageous mother, protecting her new pup from all intruders. She rebuked the marauding males patrolling the waters in hopes of mating with her. She moved her pup down the beach on several occasions to avoid the prying eyes of curious humans.

As for all Hawaiian monk seal mothers, the period of bonding was short. By the time the celebration for KP2's arrival back in the islands had become a memory, PK1 was weaned and learning to fend for himself in the waters around Kauai. Their mother had mated and was already at sea restoring her body for next year's pup in the cycle of life that was a female's destiny.

. . .

RK22, THE PURPORTED Bad Mother of Kauai, had finally successfully reared a pup. She had produced a brother for the ambassador of
their species living at the Waikiki Aquarium. The two siblings would
never meet and yet were intimately connected by their biology. One
brother would spend his life enjoying the wild waters of Hawaii while
the other would live among people to help protect the waters where
his brother played.

Acknowledgments

The idea for this book evolved during an unsettled November that began with my first meeting with KP2 at the Waikiki Aquarium and ended with an extraordinary Thanksgiving drive with him through a maze of Los Angeles highways. Proving that brilliant literary agents possess extrasensory perception, Noah Lukeman e-mailed to ask what I was up to. Clearly, big adventures were afoot. At the time, I had no idea where the journey would take the orphaned seal now in my care. Nor did I know how our story would unfold. One thing was apparent: Hawaiian monk seals were in trouble and my team would be consumed with trying to save them during the foreseeable future.

With little more than that spark, Virginia (Ginny) Smith at The Penguin Press agreed to join Noah and me on our journey with KP2. In addition to being a remarkable editor, Ginny was a welcomed, calm harbor during the stormiest periods of KP2's odyssey. With her assistant, Mally Anderson, Ginny graciously guided me through the transition from scientific writer to storyteller. She and Mally did this with the skill of experienced animal trainers, using a combination of firmness, fairness, and a biscuit reward in the end. They clearly demonstrated that an old dog can learn new tricks and I will forever be grateful for the lessons.

One could not have asked for a better crew to share KP2's oceanic message with the world. In their unique ways Noah, Ginny, and Mally have helped to save this endangered species.

It is unfortunate that a book can have only so many pages; by necessity, numerous stories and people had to be left behind. Despite being an orphan, KP2 was blessed with an extended human family that played a major part in his odyssey. Some I met, others were simply legends. All were part of this story even if not mentioned by name. They include islanders from Kauai (Mary and Barry Werthwine), Molokai (Val Bloy, Diane Pike, Julie Lopez, Uncle Walter and Aunt Loretta Ritte, Karen Holt), and Oahu (the members of the Hawaii Monk Seal Recovery Team Oahu with Donna Festa, Sharon Cosma, Jennifer Caswell, Rafe Maldonado, Dana Jones, Robert and Barbara Billand, who introduced me to wild monks, DB Dunlap, Kathy and Jim Brown, Barbara and Ralph Allen, Karen Bryan, Karen Rohter, Diane Gabriel, Stacey Stella, Lesley Macpherson, Anthony Querubin, and Dr. Gregg Levine), as well as the members of the Marine Mammal Center in Sausalito, California. The U.S. Coast Guard, the U.S. Navy (especially Mark Xitco), the U.S. Marine Corps, and the 446th Airlift Wing Air Force Reserve were KP2's military family. Rounding out his government-associated siblings were Jeff Walters, Dera Look, Trisha Kehaulani Watson, and the Maryland-based Office of Protected Resources. KP2's California family was made up of the student volunteers at the heart of my Marine Mammal Physiology Project that included Ashley Hyde-Smith, Hillary Mills, Bryan Tom, Kristen Elsmore, Courtney Roth, Megan Reis, Val Lew, Andy Garcia, Donna Beckett, Krysta Walker, Andrea Gomez, Katie Lorenz, Emily Trumbull, Hershel Krom, Christina Doll, Chelsea Aydelott, Ben Weitzman, Rachel

Tolliver, Nick Alcaraz, Caitlin Carrington, Margaret Cummings, Meagan Davis, Maia Goguen, Meghan MacGregor, Audrey Fry, Hannah Ban-Weiss, Shusuke Aihara, Hilary Walecka, Heather Tyler, Breanna Beck, Ryan Stephenson, Brett Banka, Sasha Curtis, Courtney Ribeiro-French, and many others over the years. They all worked alongside Beau and Traci, each spending long nights and even longer days caring for every need and thawing every fish for KP2. Peter Read and the Otter Cove Foundation helped to keep KP2 in fish for his two years with us and allowed me to take a risk with this endangered seal. You are all the future of this planet.

Ultimately, KP2 found a new home and a happy ending to his odyssey at the Waikiki Aquarium. I especially thank Andy Rossiter for taking the risk and Leah Kissel for taking KP2 under her training wing.

The final thank-you goes to Jim, who is on the last page but remains first in my heart (with Bart, Devon, Badger, and Kiska, of course!).

Additional Reading
and Resources

Getting Involved

By reading this book, you have already begun to help the Hawaiian monk seal. Part of the proceeds from *The Odyssey of KP2* will go directly to support Hawaiian monk seal research and public outreach at the University of California, Santa Cruz.

Additional information about supporting marine mammal conservation and research can be found at www.mmpp.ucsc.edu.

Web sites

KP2 Web site: www.monkseal.ucsc.edu.

KP2 Facebook: Hoʻailona Monk Seal (Kptwo Monk Seal).

Terrie Williams Lab: http://bio.research.ucsc.edu/people/williams.

Hawaiian Monk Seal Response Team Oahu (HMSRTO): http://hmsrto .org.

Monachus Guardian: www.monachus-guardian.org.

Endangered Species

IUCN Red List of Threatened Species (2011): http://www.iucnredlist.org.

MONK SEAL BIOLOGY AND CONSERVATION

L. R. Gerber et al. (2011). Managing for extinction? Conflicting conservation objectives in a large marine reserve. *Conservation Letters* 4(6): 417–422.

National Marine Fisheries Service (2007). Recovery Plan for the Hawaiian Monk Seal (*Monachus schauinslandi*), second revision. National Marine Fisheries Service, Silver Spring, MD. 165 pp.

Pacific Islands Region Marine Mammal Response Network Activity Update, 2008. May–August, 9.

T. M. Williams et al. (2011). "Metabolic demands of a tropical marine carnivore, the Hawaiian monk seal (*Monachus schauinslandi*): Implications for fisheries competition." *Aquatic Mammals* 37(3): 372–76.

HUMANS AND ANIMALS

M. W. Cawthron (1994). "Seal finger and mycobacterial infections of man from marine mammals: Occurrence, infection and treatment." *Conservation Advisory Science Notes* 102, Department of Conservation, Wellington, NZ. 15 pp.

J. A. Serpell (2002). "Anthropomorphism and anthropomorphic selection: Beyond the 'cute response.'" *Society and Animals* 10(4): 437–54.

MONK SEAL EVOLUTION AND ORIGINS

C. A. Fyler et al. (2005). "Historical biogeography and phylogeny of monachine seals (Pinnipedia: Phocidae) based on mitochondrial and nuclear DNA." *Journal of Biogeography* 32: 1267–79.

Homer, *Odyssey*, trans. W. Shewring. Oxford's World Classics, Book 4. New York: Oxford University Press, 1998.

W. M. Johnson et al. (2011). "Mediterranean monk seal (*Monachus monachus*)." Monachus-guardian.org.

C. B. Stringer et al. (2008). "Neanderthal exploitation of marine mammals in Gibraltar." Proceedings of the National Academy of Sciences 105(38): 14319–24.

MARINE POLLUTION AND MICRONUTRIENT CYCLING

O. J. Dameron et al. (2007). "Marine debris accumulation in the Northwestern Hawaiian Islands: An examination of rates and processes." *Marine Pollution Bulletin* 54(4): 423–33.

J. R. Henderson (2001). "A pre- and post-MARPOL Annex V summary of Hawaiian monk seal entanglements and marine debris accumulation in the Northwestern Hawaiian Islands, 1982–1998." *Marine Pollution Bulletin* 42(7): 584–89.

T. J. Lavery (2010). "Iron defecation by sperm whales stimulates carbon export in the Southern Ocean." *Proceedings of the Royal Society*, B doi:10.1098/rspb.2010.0863.

Index